Mapping Wilderness

Stephen J. Carver • Steffen Fritz
Editors

Mapping Wilderness

Concepts, Techniques and Applications

Springer

Editors
Stephen J. Carver
School of Geography
University of Leeds
Leeds, UK

Steffen Fritz
Ecosystem Services and Management
International Institute of Applied
System Analysis
Laxenburg, Austria

ISBN 978-94-017-7397-3 ISBN 978-94-017-7399-7 (eBook)
DOI 10.1007/978-94-017-7399-7

Library of Congress Control Number: 2016931854

Springer Dordrecht Heidelberg New York London

Printed on acid-free paper

Springer Science+Business Media B.V. Dordrecht is part of Springer Science+Business Media (www.springer.com)

Robert George Lesslie

This book is dedicated to the memory of Rob Lesslie, who sadly died in March 2014 during the writing of this book.

Rob was a leading Australian geographer and ecologist whose working career spanned more than 30 years in natural resources evaluation and management in government, education, and the private sector. Throughout his career he published numerous book chapters, journal papers, and many technical reports, though Rob's impact went well beyond this, with his being significantly involved in some of the most pressing natural resource management policies throughout his extensive career.

Rob was instrumental in developing and delivering the Australian Government's National Wilderness Inventory programme (1986–1996) and more recently led an advisory team that developed legislation for wilderness protection in South Australia, the Wilderness Protection Act. Currently some 1.8 million hectares of wilderness in SA is protected under the Act as a result of a state-wide assessment based on Rob's work.

Rob played a significant role in several initiatives: shaping the Australia-China Environment Development Program (ACEDP), developing the scientific framework for the WildCountry project, and initiating the development and ongoing enhancements and applications of the Multi-Criteria Analysis Shell for Spatial Decision Support. Rob was also the driving force between the establishment of a memorandum of understanding between ABARES and the Chinese Forestry Economics and Development Research Centre to support future cooperative research.

Of particular relevance to this book is how Rob's early work on mapping wildness has influenced many of those who followed, including both of the book's editors. We both met Rob on separate occasions and were touched by his open friendship, generosity, gentle encouragement, and willingness to share his ideas and experience. Rob's work developed an important foundation for the mapping, identification and management of the world's wilderness areas that are considered in this book.

Preface

Wilderness mapping may sound like a bit of an esoteric topic, but when you think carefully about it, it ought to be a fundamental theme within environmental science and management. Wilderness is in many ways the pristine resource. It is where we as humans have historically derived all of our goods, services, and resources, save perhaps for human ingenuity itself – though one might argue that the development of even that has its origins in our distant wild past wherein survival and development as a species depended on intelligence and invention. Even today in our advanced technological society, we derive so much from wilderness and wild landscapes that we often take for granted. Clean air and water are the most obvious, but we must not forget the other regulating and supporting services provided by wilder lands including flood retention, carbon storage and sequestration, wildlife habitats, and recreational environments. Wilderness also provides us with subjects of scientific study where we can observe natural processes and wildlife free from human interference. We should also acknowledge their importance as a source of not only scientific knowledge but also artistic and personal inspiration.

Today, wilderness is a shrinking resource. There are few areas on the planet that we have not explored or exploited to some degree or another. Even the world's last great wilderness areas are known, mapped, and studied. Indeed, many show signs of irreversible impacts from climate change and global pollution. This is only likely to increase in the future as population growth generates ever-increasing demands for land and resources. Mapping what is left and using the knowledge and information thus derived are therefore essential if we are to protect and preserve wilderness in the future for the benefit of both humans and wildlife.

This book has been a bit of a slow train coming. It was several years ago now that a chance conversation between one of the editors and the publishers suggested that a PhD could be turned into a book. That in turn changed into the idea for a volume on the broader topic of wilderness mapping, which then in turn morphed into the edited volume you now have before you. The book attempts to give a comprehensive overview of the topic area covering the conceptual and philosophical foundations, techniques and methodological approaches, and applications at a variety of spatial scales. In doing so we have brought together a range of contributors who are both experts in their field and cutting-edge thinkers in the wilderness and spatial mapping domain. Spatial information technology and mapping science is a rapidly expanding and developing field, and so we expect to be able to add to this in the future. For now, the book provides a record of the 'state of the art' and will, it is hoped, enable you the reader to follow our lead and map your wilderness.

Leeds, UK
Laxenburg, Austria
Spring, 2015

Stephen J. Carver
Steffen Fritz

Acknowledgments

There are many people we ought to acknowledge in the journey of this book from idea to reality. Firstly, we would like to thank Margaret Deignan from Springer for her initial suggestion and then her patience and tenacity in seeing the book through to completion. Margaret and her team at Springer including Takeesha Moerland-Torpey and Corina van der Giessen have been helpful all the way, not least in being relaxed about deadlines that came and went. Secondly, we would also like to thank Rachael Unsworth for volunteering – without being asked – to proofread all the text which she has done expertly and with such a careful eye for detail. There are many others deserving of thanks, but from the editors to our authors, again we would like to thank you for your patience and hope the wait to see your work in print has been worth it.

Contents

About the Authors

Kees Bastmeijer Kees is professor of nature conservation and water law at Tilburg Law School (Tilburg University, the Netherlands). His research focuses on the role of international, European, and domestic law in protecting nature. His research themes include nature conservation in the polar regions, philosophical human-nature attitudes and the functioning of nature protection law, property rights and nature, the implementation of the EU Natura 2000 regime, and the role of law in protecting wilderness.

Bill Borrie Bill has a PhD from Virginia Tech but has been with the University of Montana since 1995. He has conducted research in protected areas such as the Boundary Waters Canoe Area Wilderness, Yellowstone National Park, the Bob Marshall Wilderness Complex, the Arctic National Wildlife Refuge, Okefenokee National Wildlife Refuge, and Petrified Forest National Park. His research interests are focused on the meanings and experiences of parks and wilderness.

Stephen J. Carver Steve is a geographer with an interest in the application of GIS to wilderness and wild landscapes. He is particularly interested in quality assessment, public participation, and visibility analyses. He has worked widely on these topics in the UK, Europe, and North America. He is director of the Wildland Research Institute and is based in the School of Geography, University of Leeds.

Roger D.J. Catchpole Roger is an ecologist with an interest in how landscape structure and climate change affect the dispersal processes of plants and animals. He has worked widely across the UK and Europe as an expert advisor in this field. He has also applied spatially explicit models to help target protected area conservation effort at a national level and is an experienced capacity builder. He is an honorary fellow in the School of Geosciences at the University of Edinburgh and an active botanist.

Mark L. Douglas Mark is assistant professor of Environmental Recreation and Tourism Management at the University of Maine at Machias. He teaches, researches, and works in recreation resource management and conservation social science. His focus is on the role that stewardship plays in society and culture. At press, he is developing his theory of wildhood, the elemental kinship of the wild in people, places, and things.

Karin Enzenhofer Being a vegetation ecologist by training, Karin has her technical basis on the topic ecology of forest ecosystems with special interest on hemeroby. Working for WWF for over 2 years, she has been contributing to projects of sustainable land-use planning, connectivity, and wilderness in the Alps with special focus on Geographic Information Systems.

Jonathan Erickson Jonathan earned a master of science in conservation social science from the University of Idaho in 2008. His graduate research focused on a recreation economic analysis of the Lower Salmon River for the Bureau of Land Management. Currently, Jonathan works for the Columbia River Gorge National Scenic Area. He has worked closely on a number of monitoring programmes that utilised GIS data collection methods including recreation site analysis, social trail monitoring, and winter law enforcement tracking.

Justin Ewer Justin currently works as a wilderness manager on the Gifford Pinchot National Forest. His previous experience includes work in several other wilderness areas across Washington and Oregon. He has led several wilderness data collection projects and incorporated GIS technology to improve the efficiency and data analysis capabilities of these projects. Working for the USDA Forest Service has also given him the privilege of fighting fires, wrestling wolverines, building trails, and restoring damaged landscapes.

Steffen Fritz Steffen is head of the Earth Observation Systems group in the Ecosystems Services and Management Program at IIASA. Steffen is the main driver behind Geo-Wiki and is currently researching different incentive schemes for crowdsourcing of land cover in Kenya, Tanzania, and Austria through the European Research Council-funded project called CrowdLand. His PhD was on wilderness mapping under the supervision of Stephen J. Carver.

Adrian Hăgătiş Adrian is a geographer and team leader of the Southern Carpathians programme – Rewilding Europe and WWF Danube-Carpathian Programme. He is involved in the management of the projects that aim to restore biodiversity in the area and to establish wilderness areas in the Southern Carpathians. He has grown up close to the Carpathians and studied them from an early age. His passion for nature conservation has become a way of life, inspiring others to help in his quest.

Benjamin D. Hennig Ben is an academic geographer educated at the Universities of Cologne and Bonn and the Alfred Wegener Institute for Polar and Marine Research (Bremerhaven, Germany) where he conducted research on hyperspectral remote sensing applications in coastal ecosystems. He completed his PhD as part of the Worldmapper project at the University of Sheffield (UK) with research on visualising the social dimensions of our planet. He now works as a senior research fellow in the School of Geography and the Environment at the University of Oxford.

Flora Hoser Flora is an ecologist by training and has been working for WWF since 2010 in the fields environmental education, species conservation programme for raptors, as well as protected area management, especially in the Pannonian lowlands of Eastern Austria.

Bernhard Kohler Bernhard is a biologist and currently head of the Biodiversity Programme of WWF Austria, as such involved in WWF Austria's wilderness focus since 2010. Bernhard's special interests are the identification, design, and promotion of wilderness areas, the preservation of old growth attributes in managed forests, and the management of Pannonian wetlands.

Rob Lesslie Rob was a geographer and ecologist with interests in natural resources evaluation and management in government, education, and the private sector. He was instrumental in developing and delivering the Australian Government's National Wilderness Inventory programme (1986–1996) and more recently led an advisory team that developed legislation for wilderness protection in South Australia, the Wilderness Protection Act. Sadly Rob passed away during the production of this book.

Lisa K. Machnik Lisa is the Pacific Northwest Wilderness/Wild and Scenic Rivers/Congressionally Designated Areas programme manager for the USDA Forest Service. An avid backpacker, canoeist, trail runner, and skier, Lisa worked as a backcountry guide in Canada and New Zealand before completing her PhD as the inaugural Hartzog fellow at Clemson University. At Clemson, she worked with the National Park Service in their Interpretive Development Program before joining the Forest Service as a Presidential Management Fellow.

Dragoş Ştefan Măntoiu Dragoş is a geographer, specialised in the field of GIS and biodiversity for research and consultancy purposes. Dragoş's main interests are bat ecology with focus on migration routes and wind farm impact, wilderness mapping, and species distribution modelling issues. A nature enthusiast, Dragoş is always willing to lend a helping hand regarding nature conservation projects or studies which may benefit from a GIS analysis. He is coordinating a conservation NGO (Wilderness Research and Conservation) created to integrate studies related to wild land mapping.

Ionuţ Cornel Mirea Ionuţ is a geographer, specialised in the field of GIS and geology, currently working at the Institute of Speleology 'Emil Racoviţă' in Bucharest. His main area of expertise includes palaeontology studies, cave surveying, and environmental modelling. His passion for speleology and wilderness inspired students within the field.

Marius Costin Nistorescu Marius is an ecologist, specialised in the field of environmental impact assessment and strategic impact assessment, and has devoted his career towards the consultancy sector, where he coordinates important studies regarding the environmental impact upon biodiversity and social issues, at a company called EPC Consultanţă de mediu, in Bucharest. Marius is now coordinating a team which will propose a first regional strategy for wilderness management in Romania.

Rannveig Ólafsdóttir Rannveig is a physical geographer with her main focus on the interrelationship between tourism and the environment. Rannveig works as a professor in tourism studies at the Faculty of Life and Environmental Sciences, University of Iceland. Her special research interests are on tourism environmental impacts, tourism and wilderness, wilderness mapping, geotourism, tourism spatio-temporal modelling using system dynamics and GIS, public participation, and sustainable tourism management.

Christoph Perger Christoph is the lead Geo-Wiki developer, having designed the first version of this tool for his master's dissertation at the University of Applied Sciences in Wiener Neustadt. He continues to develop Geo-Wiki and the latest online validation tool called LACO-Wiki, at IIASA. He has developed numerous Geo-Wiki branches, e.g. to validate Australian land cover, a cities branch for collection of urban properties and a pictures branch for displaying geo-tagged photos from the Pictures Geo-Wiki mobile application.

Christoph Plutzar Christoph is biologist and social ecologist, and he holds an academic training in Geographic Information Systems. Since 2008 he is senior researcher at the Institute of Social Ecology (Alpen-Adria-University Klagenfurt-Wien-Graz), working in the area 'Land Use and Colonization of Ecosystems'. The focus of his work is on handling spatially explicit research questions with GIS, especially in the topics land use, biodiversity, wilderness, and nature conservation.

Micael Runnström Micael is a physical geographer and works as a senior lecturer at the Department of Physical Geography and Ecosystem Science at Lund University, Sweden. His main research and teaching areas are applied methodologies in GIS and remote sensing techniques. His main research interest is on environmental change in respect to anthropogenic impacts and use.

Ionuţ Cosmin Şandric Ionuţ is a geographer and lector at the Faculty of Geography, University of Bucharest, and has great expertise in the field of statistics, environmental modelling, remote sensing, and GIS solutions, including online and open source sectors. Ionuţ's main area of interest revolves around natural hazards, but has coordinated MScs thesis on a wide range of subjects, including biodiversity. His work also includes consultancy for ESRI Romania.

Neil Sang Neil is a researcher in the Department of Landscape Architecture, Planning and Management at the Swedish University of Agricultural Sciences at Alnarp near Lund and runs AnserGIS.com. His research interests include visual landscape modelling, landscape perception, land abandonment modelling, spatial data structures, MAUP, citizen science, GeoDesign, spatial statistics, SDI and Integrated Environmental Modelling. Previous to moving to Sweden, Neil worked at the James Hutton Institute in Aberdeen.

Anna Dóra Sæþórsdóttir Anna Dóra is a geographer with interest in nature-based tourism, wilderness tourism, tourism management, and the land-use conflicts between tourism and other sectors, for example, the power production industry. She is an associate professor in tourism studies at the Faculty of Life and Environmental Sciences, at the University of Iceland.

Linda See Linda is a research scholar at the International Institute for Applied Systems Analysis (IIASA) in Austria. She works in the Geo-Wiki team on the development of crowdsourcing campaigns, the analysis of the data, and on the development of hybrid land cover maps that use crowdsourced information.

Erika Stanciu Erika is a forestry specialist and one of the most important speakers regarding nature conservation in Romania and Europe; former president of EUROPARC, the umbrella organisation for Europe's protected areas; and also the executive director at the ProPark Foundation in Romania, an organisation that helps train specialist in the field of nature conservation. She was also the head of WWF's Carpathians/Forests and Protected Areas Programme. Her work inspired many followers towards the path of nature conservation, yielding a new work sector within Romania.

Michael Zika Michael is an ecologist with a focus on human influence on ecosystems. He is working for WWF for over 6 years on conservation projects in the Alps and the Greater Mekong region and on nature conservation policy. His special interests are on wilderness, landscape integrity, PAs, and sustainable land-use practices, linking biodiversity protection and footprint reduction. From 2015 she holds the position of state secretary affiliated to the Ministry of Environment, Waters and Forests (MMAP). During the past years, he also developed or contributed to projects on sustainable land-use planning, ecosystem restoration, connectivity, and ecosystem service valuation.

Chapter 1
Introduction

Stephen J. Carver and Steffen Fritz

Abstract Wilderness and wilderness definitions are complex and problematic and therefore present particular difficulties for mapping and GIS, both of which depend largely on carefully defined attributes and discrete criteria. The rationale for mapping wilderness is described and our interest in the topic justified in terms of wilderness protection, conservation, human benefits and nature. The threats to wilderness are legion and somewhat obvious to anyone with even a basic understanding of the planet. Human population growth and associated demand for land, food and resources is the key impact on wilderness. Road construction opens up wilderness areas for exploitation, farming and settlement. Even our attempts to lessen our impact on global climate and ecosystems can lead to further reductions in wildness (e.g. renewable energy technologies while reducing our carbon footprint can have marked impacts on wild landscapes). The basic concepts of wilderness mapping are outlined and a brief history of wilderness mapping described including key developments at global, regional and local scales. The structure and contents of the book are given.

Keywords Wilderness • Definition • Mapping

1.1 Towards a (Spatial) Definition of Wilderness

> One man's wilderness is another's roadside picnic ground (Nash 1993, p. 1)

As definitions of wilderness go, this is perhaps both helpful and problematic in equal measure. It is helpful in that it underlines the vagueness of the concept and the fact that different people, with different backgrounds and life experiences, will perceive wild landscapes in different ways. Ultimately it is our familiarity with the wilderness condition that will determine where on a scale of human modification from most to least modified that

S.J. Carver (✉)
School of Geography, University of Leeds, Leeds, UK
e-mail: s.j.carver@leeds.ac.uk

S. Fritz
Ecosystem Services and Management, International Institute of Applied System Analysis, Laxenburg, Austria
e-mail: fritz@iiasa.ac.at

S.J. Carver, S. Fritz (eds.), *Mapping Wilderness*,
DOI 10.1007/978-94-017-7399-7_1

we feel wilderness begins and ends. Conversely, Nash's definition is also problematic in that it gives us nothing by way of any kind of "yardstick" or definitive criteria against which we can measure, and therefore map, this thing we call wilderness.

This book focuses uniquely on the approaches, techniques and attempts to map and model wilderness from a geographical perspective at a range of spatial scales covering a variety areas. There is a steadily increasing literature on wilderness mapping that attempts in a rigorous, robust and repeatable manner to say exactly what we are talking about when we speak of wild places and map where they are, such that we can best manage our influence on them and design policies for their protection. In this respect, Nash's definition actually isn't a bad place to start as it succinctly, in just one sentence, points simultaneously to both the problem and the solution. The problem is that there is no single accepted definition of what wilderness is (and isn't) and that it depends very much on the point of view of the individual. As such, the solution to what wilderness is and where it can be found or said to exist is, at least from a spatial science perspective, a classic ill-defined and fuzzy multi-criteria spatial concept and can be largely addressed using existing methods and tools.

Nash's "one man's wilderness" definition is a sociological one. While it is philosophically interesting and points towards a solution, it is hardly a useful legal definition nor is it all that helpful in tight geographical terms. Other definitions have been developed, however, developing over time in sophistication and clarity. Some present a better set of indicators that lend themselves to being mapped. Perhaps the best known of which is that which accompanies the 1964 US Wilderness Act.

> A wilderness, in contrast with those areas where man and his own works dominate the landscape, is hereby recognized as an area where the earth and its community of life are untrammeled by man, where man himself is a visitor who does not remain. An area of wilderness is further defined to mean in this Act an area of undeveloped Federal land retaining its primeval character and influence, without permanent improvements or human habitation, which is protected and managed so as to preserve its natural conditions and which (1) generally appears to have been affected primarily by the forces of nature, with the imprint of man's work substantially unnoticeable; (2) has outstanding opportunities for solitude or a primitive and unconfined type of recreation; (3) has at least five thousand acres of land or is of sufficient size as to make practicable its preservation and use in an unimpaired condition; and (4) may also contain ecological, geological, or other features of scientific, educational, scenic, or historical value. The Wilderness Act, Public Law 88–577 (16 U.S. C. 1131–1136) 88th Congress, Second Session September 3, 1964

The Wilderness Act talks of absence of human artefacts and management, lack of human settlement, remoteness, opportunity for solitude, ecological condition and size. All of these criteria can to a greater or lesser extent be mapped using modern digital datasets and computer software. It has also been influential in expanding wilderness definition and protection worldwide. The IUCN now defines wilderness under Category 1 either as Category 1a (Strict Nature Reserve), which are areas set aside primarily for research, or Category 1b (Wilderness) defined as:

> Large areas of unmodified or slightly modified land and/or sea, retaining its natural character and influence, which is protected and managed so as to preserve its natural condition… [wherein the primary objectives are] To protect the long-term ecological integrity of natural areas that are undisturbed by significant human activity, free of modern infrastructure and where natural forces and processes predominate, so that current and future generations have the opportunity to experience such areas. (IUCN 2008).

Looking at this definition and comparing it to the text from the US Wilderness Act (1964) it is easy to see where the inspiration for the IUCN definition comes from. The language and wording may be different but the message and meaning is exactly the same.

The following year a European Parliament Resolution on Wilderness called for the development of guidelines on managing wilderness within Natura 2000,[1] a unified European definition of wilderness and a register of remaining areas. The document "*Guidelines on Wilderness in Natura 2000: Management of terrestrial wilderness and wild areas within the Natura 2000 network*" was published in 2013 and contains the following definition:

> A wilderness is an area governed by natural processes. It is composed of native habitats and species, and large enough for the effective ecological functioning of natural processes. It is unmodified or only slightly modified and without intrusive or extractive human activity, settlements, infrastructure or visual disturbance. (EC 2013, p. 10)

Again, like the US Wilderness Act that precedes them, spatial criteria or attributes of wilderness are evident within the text of these definitions. This then begs the question as to how we can translate these definitions into reliable maps that would be useful in policy-making and decision support roles?

1.2 Why Wilderness?

While definitions are usually a good place to start any book on a particular subject, it is also useful to identify just why it is worth studying a particular topic. Wilderness is in many ways the ultimate and pristine resource. It is the very stuff from which we have made the human world. It is where we have come from and it may be where we are going. It is where we have carved out civilisations and cultures, and drawn our resources of land, water, oil, gas, minerals, timber, fish and game, etc. Over the millennia humans have pretty much exploited, with only a few exceptions, every last ecosystem of the planet. We have cut down forests, ploughed up the land and built huge cities connected by dense and efficient networks transporting people, goods, resources and information. In the process we have greatly modified whole landscapes and ecosystems and have easily become the dominant species on the planet, making thousands, perhaps millions, of others extinct in the process.

The US Wilderness Act came into being as a result of a gradual realisation over the preceding years that the frontier was fast disappearing and that something needed to be done to preserve America's last wilderness areas and the wildlife that depends on them. This was significant as the frontier is arguably what made America (notwithstanding the indigenous native population) and by 1964 there was no longer a frontier in the lower forty eight. A long-running campaign of lobbying for and promotion of the wilderness ideal preceded the signing of the Wilderness Act on 3rd September 1964 by President Lyndon B Johnson. In fact it took Howard Zahniser no less than 15 years to draft the text of the Act and see it through Senate.[2] The list

[1] Natura 2000 is the pan European nature protection network.

[2] Zahniser tragically died just a few months before the final signing of the Act.

of those involved in the run up to this moment reads something like a "who's who" roll call of the most famous names in wilderness advocacy… John Muir, Aldo Leopold, Sigurd Olsen, Arthur Carhart, Bob Marshall, to name but a few. Their concern for the loss of wilderness as the frontier receded was driven variously by their own need or desire for wild spaces to be wild in …

> Man always kills the thing he loves, and so we the pioneers have killed our wilderness. Some say we had to. Be that as it may, I am glad I shall never be young without wild country to be young in. Of what avail are forty freedoms without a blank spot on the map? (Aldo Leopold 1949)

… but also out of a realisation that the loss of wilderness also meant something much, much more. Wilderness represents more than just landscapes empty of human endeavour. Wilderness itself is important for free flowing rivers and for the clean water supplied to nearby conurbations. Wilderness provides a habitat and refuge for wildlife. Wilderness areas are important for science, providing as they do control environments against which we can gauge, measure and monitor our impact on the natural world. In today's language we might call these ecosystem services. De Groot et al. (2002) split these into four types: provisioning, regulating, supporting and cultural. Table 1.1 gives examples of these services and how these are best provided and modulated in wilderness as opposed to human modified ecosystems.

One "service" that doesn't fit easily into such a classification – for the classification itself always stresses the anthropogenic benefits – is the intrinsic value of wilderness. That is to say, do natural processes, landscapes, species and the ecosystems they represent (i.e. nature herself[3]) have to be commodified to have value, and does

Table 1.1 Ecosystem services and wilderness

Service class	Traditional/extractive	Sustainable/non-extractive
Provisioning	Timber, (bush)meat and other foodstuffs, fibre/furs, minerals, oil and gas, renewable energy	Clean water,[a] carbon storage, genetic material, clean air
Regulating		River flows, erosion control and influence on sediment yields, nutrient supply, carbon sequestration, pollution stripping, climate regulation
Supporting		Natural cycles (e.g. hydrological cycle, carbon cycle, nitrogen cycle, etc.), wildlife habitats, climate systems
Cultural	Hunting and fishing grounds, wildlife herding/harvesting, timber harvesting, collecting foodstuffs (fungi, berries, plant material, etc.)	Recreational landscapes, wildlife observation, landscape aesthetics/appreciation, artistic inspiration, cultural heritage,[b] intrinsic values

[a]Abstraction from rivers flowing out of wilderness areas or ground water abstraction tapping into resources recharged within wilderness areas
[b]Often evidenced as archaeological remains or written/oral histories and legends

[3]The origin of the word "Nature" is from the Latin *natura* meaning "birth" and thus gives rise to early representations of nature as female and the popular image of "Mother Earth".

that value have to accrue in any tangible way to us as humans? Wilderness, like nature in "raw" form (for that is what wilderness surely is in the final analysis) has value beyond human needs for all other species we share the planet with. Contrary to the flawed reasoning of the modern green movement whose positivist approach maintains that human dominion is natural in of itself and we can engineer, plan and design our way out of ecological disaster, wilderness is a necessity for planetary survival. This was recognised as early as 1862 when Henry David Thoreau penned the immortal words *"In wildness is the preservation of the world"*.

Without the biophysical diversity that characterises intact ecosystems and the natural processes that drive these, we are ultimately doomed, for life on Earth depends on these provisioning, supporting and regulating services to make the planet habitable. If policy-makers, planners and commercial enterprise feel more comfortable with financial devices and arguments, then it has been calculated that the total annual economic worth of the natural environment to the global economy is in the region of $44 trillion, or roughly twice that of global GNP (Costanza et al. 1997). Of course, there will be some that say these benefits will still accrue regardless of whether there is wilderness or not, but it seems a safer bet that wilderness ecosystems taken as a whole provide a far greater economic benefit in terms of their ecosystem services than do the equivalent area of human modified systems.

A new movement was launched at the end of 9th World Wilderness Congress (WILD9) held in Merida, Mexico in 2009. This was Nature Needs Half (NNH), the central tenet of which is that we should aim to protect at least half of the world (both terrestrial and marine) for nature. The basic concept here is an ethical argument that reasons that if humans manage half the planet for agriculture, industry and settlement, and the other half is devoted to nature conservation then this provides a reasonable basis for a sustainable planet. NNH *"recognizes that human well-being and security depend greatly on a healthy, resilient, and abundant natural world... and also that Nature itself has a right to exist freely"*. Of course the key question is "where?" as well as what ecosystems might reasonably be represented? These are spatial questions to which the methods, techniques and approaches in this book might well be applied.

1.3 Mapping the Wild

There are a number of existing mapping projects that have attempted to illustrate exactly where the world's wilderness areas are and the ecosystems represented therein. The first such project was a global reconnaissance of wilderness areas carried out by McCloskey and Spalding for the 4th World Wilderness Congress in 1987. The subsequent paper in Ambio published a figure suggesting 34 % of the world's land area could be classified as wilderness (McCloskey and Spalding 1989). This was a simple rule-based map for which they used two principal Boolean criteria: areas more than 400,000 ha in extent and greater than 6 km from any recorded human feature – as based on data derived from Jet Navigation Charts at a scale of 1:2 million. This was a remarkable feat given that all the work was done by hand with paper maps before GIS was mainstream technology.

1.4 Here Be Dragons

Although the McCloskey and Spalding map was the first coordinated attempt at mapping global wilderness we have arguably been doing it for hundreds of years, not by dint of what we have mapped, but rather what we haven't. Take a close look at any old map of the world produced before around 1800 and invariably you'll find examples of Leopold's blank spots. These are sometimes labelled "Parts unknown" indicating the cartographer had no knowledge or information as to what lay over the horizons of the known world with which to fill the white space on the page. In some older maps cartographers often used their imagination and filled in the empty spaces with flights of fancy including imaginary lands and seas inhabited by strange and wonderful beasts and equally wild and savage people. The Latin phrase "HC SVNT DRACONES" (meaning literally "here be dragons") was sometimes used to indicate such wild and fearful places. Even as recently as the mid 1700s, navigation charts of the north Atlantic eluded to the presence of Buss Island, an uncharted island that has since been proved not to exist. The advent of aerial photography and earth observation satellites in the twentieth century means that every inch of the planet is now mapped and "known" even if, as in some parts of Antarctica for example, we can be confident that no human has ever set foot there and so remain inviolate. Of course, this view of the "known world" is a particularly Eurocentric one and we acknowledge that nearly all lands were already discovered, and therefore known, by indigenous native populations long before Europeans arrived to chart their existence and exploit their bounties.

1.4.1 Digital Worlds

Nevertheless, the McCloskey and Spalding map marks the start of a period of intensive mapping activity across the globe aimed at mapping human impact and the last wild places. Arguably the first proper use of GIS to map wilderness quality was the Australian National Wilderness Inventory (NWI) as described by Rob Lesslie in the following chapter of this book (see Fig. 2.2). Here national digital datasets are used within a cartographic model to rate wilderness quality based on four criteria: remoteness from mechanised access, remoteness from settlement, apparent naturalness (distance from modern human artefacts) and biophysical naturalness (naturalness of the land cover) (Lesslie and Maslen 1995). These two basic factors – remoteness and naturalness – are used in one form or another in nearly all models of wilderness quality. Lesslie (1998) expanded the NWI concept to the whole world in work done for the World Conservation Monitoring Centre (WCMC) (see Fig. 2.3). Eric Sanderson and his team at Wildlife Conservation Society (WCS) and the Columbia University Center for International Earth Science Information Network (CIESIN) adapted this approach using multi-criteria mapping techniques to create a global map of the Human Footprint showing the degree of human influence according to nine global data layers covering human population pressure, human land use and infrastructure, and human access (Sanderson et al. 2002). This map is then used to map The Last of the Wild based on an interpretation of the Human Footprint data within global biomes.

Other mapping programmes have followed a similar approach. The Globio map accompanying the UN Global Biodiversity Outlook programme is a good global scale example, while regional and local scale mapping follows a similar model though often making use of more complex mapping tools that the opportunity of higher resolution datasets and smaller areas afford (see Chap. 5). Two country-level maps and a local scale map are presented here in this book for Iceland (see Chap. 11 and Fig. 11.1), Austria (see Chap. 12 and Fig. 12.1) and the Carpathian Mountains in Romania (see Chap. 10 and Fig. 10.3).

Some areas are mapped across multiple spatial scales and at varying levels of detail. Perhaps the best mapped area in terms wilderness quality is Scotland which has been mapped at a global, continental, national and local scale. The methods used to do this are essentially the same, but vary the indicators, data and models used to best suit the scale in question. The maps from the Human Footprint, Globio and WCMC place Scotland in a global context. At this scale, Scotland doesn't appear to contribute anything to global wilderness. Looking closer, a new European scale wilderness quality map has recently been developed for the European Environment Agency (EEA) based on naturalness of vegetation (measured as departure from the potential natural vegetation in the absence of human modification), and remoteness from roads and settlement at a 1 km^2 resolution (Kuiters et al. 2013). At this scale, parts of Scotland do appear to figure in the top 5 and 10 % wildest areas in Europe. Parallel to the European mapping, Scotland has produced its own wild land map at a resolution of 50 m. This is being used to directly inform Scottish planning policy and decision-making on development (SNH 2014). This was a three stage process, with a wildness continuum map (Phase 1) based on measures of naturalness of land cover, absence of modern human artefacts, ruggedness and remoteness, being used to identify core wild land areas based on a statistical reclassification of the continuum (Phase 2) and a final drawing of wild land area boundaries using information from stage 2 in coordination with local knowledge and on-the-ground boundaries such as rivers and ridge lines (Phase 3). Further detail within the Scottish national parks is provided by local level mapping at even finer resolution of 20 m. Both national parks – the Cairngorm National Park and the Loch Lomond and The Trossachs National Park – were mapped to inform developing national park planning policies on wild land (Carver et al. 2012) and acted as a feasibility study and methodological template for the national map.

A similar approach to the Human Footprint mapping has been developed for the world's seas and oceans in work by Ben Halpern's team at the National Center for Ecological Analysis and Synthesis, Santa Barbara (Halpern et al. 2008). Here no less than seventeen different datasets covering human impacts on marine ecosystems from fishing, climate change, and pollution are combined to produce an overall score of vulnerability to human activities. Marine wilderness areas remain comparatively under studied, perhaps as a result of a paucity of good quality data and the need for different spatial models that take the 2D surface and 3D submarine nature of "seascapes" into account.

A recent departure from the largely multi-criteria based work is the Roadless Areas map produced using Google Map road data which again echoes Leopold's quote about the blank spots on the map. This makes the simple assumption that virtually all human impact on the world's land area is associated with road construction and therefore a map showing distance from nearest road makes for a very good indicator of probable wilderness quality. Experience shows that where road construction takes place, people and development generally follow. A good example is where oil, gas and mineral exploration roads built into virgin forest have, in themselves, a very limited footprint but then provide easy access for logging. Over time agriculture and settlement usually follow. Thus, the impact of human development can be seen not only in a spatial but also a temporal framework.

1.4.2 Patterns, Drivers and Threats

At a global scale work by Erle Ellis and his team in the Laboratory for Anthropogenic Landscape Ecology at the University of Maryland have used historical data on population density and land use to map long-term anthropogenic changes in the terrestrial biosphere, as compared to the Potential Natural Vegetation (PNV) in the absence of humans, over a period of 300 years from 1700 to 2000 (Ellis et al. 2010). The key findings show that while in 1700 just under half the world's land surface area was wilderness (wildwood and wild treeless barren land) only 5 % could be described as intensively used. Since 1700 wildlands have reduced to only 23 % with the rest being used (40 %) and novel (i.e. anthropogenically created or modified) ecosystems (37 %). The rapid growth in the total human population, which reached seven billion people in 2012 and is projected to reach between 9 and 12 billion by 2100, is the obvious driver in terms of demand for land (for agriculture and living space) and resources meeting our ever increasing demand for goods and services. This generates an ever-increasing threat to the world's remaining wilderness areas. However, as much of the global scale mapping work shows, these remaining wilderness areas are primarily in the Earth's coldest and driest regions and so likely to show some resilience to settlement and agriculture, though not resource extraction pressures as higher prices and greater demand mean "hard to get at" resources in remote locations become economically viable, witness the recent expansion in mineral exploration in Greenland (Schønwandt and Dawes 1993).

Population growth and road construction may be the main drivers at the global scale, but subtly different forces may influence trends at regional and local scales. Work at the United Nations Environment Programme (UNEP) Arendal site in Norway provides a nice example of how road construction, urbanisation and other human infrastructure have markedly reduced the remaining areas of undeveloped land in Norway over the last 100 years (Brun 1986; Grid Arendal 1992). Carver and Wrightham (2003) show how wild land areas in Scotland have reduced over the last

150 years driven by expansion and upgrades to the road and rail network, but also from plantation forestry operations and exploitation of renewable energy sources. Initially impacts from renewables were driven by the building of hydroelectric schemes, but recently the main threat can seen as coming from large scale industrial wind farms (Carver and Markieta 2012). This is a classic "green-on-green" impact with renewable energy ostensibly trying to reduce our carbon footprint while having a damaging effect on carbon stores, ecology and amenity values (Drewitt and Langston 2006; Smith et al. 2014; Warren et al. 2005). Work by SNH has shown that the area without visual influence from built development fell from 41 to 31 % between 2002 and 2008 and then to just 28 % by the end of 2009. Much of this is attributable to wind farm development.[4] Other research within the Scottish national parks has shown plantation forestry, hill track construction and renewable energy developments to be the key impacts on wild land quality in these areas (Carver et al. 2012).

While the most obvious problem arising from the gradual, sometimes rapid, attrition of wilderness over time is the shrinkage in total area, it does come with a series of associated problems that will be familiar to anyone with a background in spatial ecology. These include fragmentation and isolation. The general pattern is directional, well-known and largely predictable, though the rate and exact spatial pattern is more difficult to predict. Given a "blank canvas" of pristine wilderness human development will occur in patches usually around the perimeter, thus generating "holes" the canvas. This is the "frontier" state wherein there is still more wilderness than developed land. Over time settlement and cleared land gradually erode the undeveloped wilderness areas and ground transportation links built to connect settlements start to fragment and divide up the hitherto contiguous wilderness area into separate core areas. There is now more developed land than wilderness giving rise to a "torn" canvas (Tin and Carver 2014). As development progresses settlements and cleared land begins to merge, assisted by expanding transport networks, and the remaining wilderness shrinks to a few core areas resulting in isolation. This process is shown in Figs. 1.1a, 1.1b, 1.1c, and 1.1d.

Over much of the developed world we have already reached the state shown in Fig. 1.1d. Many small and highly developed nations have no real wilderness areas left at all and haven't had for centuries. This is true for much of central, western and southern Europe. Where wilderness areas do remain they tend to be small and isolated. The larger the country, the greater the opportunity for intact wilderness areas to remain. The USA is an interesting example. The lower 48 states of the conterminous US has seen wilderness largely reduced to a few pockets, mainly in the west (the "torn" canvas shown in Fig. 1.1d). Much of the remaining wilderness is found in Alaska, which is arguably still a frontier state and an example of a "holey" canvas (Fig. 1.1b). Nevertheless, the National Wilderness Preservation System that arose from the 1964 Wilderness Act now protects over 750 areas totalling over 109 million acres (44.3 million hectares) of wilderness across the US of which near half of

[4] Scottish Natural Heritage (2013) Natural Heritage Indicator: N3 Visual Influence of built development http://www.snh.gov.uk/docs/B551051.pdf

Fig. 1.1a Patterns of human development in wilderness lands: Blank canvas

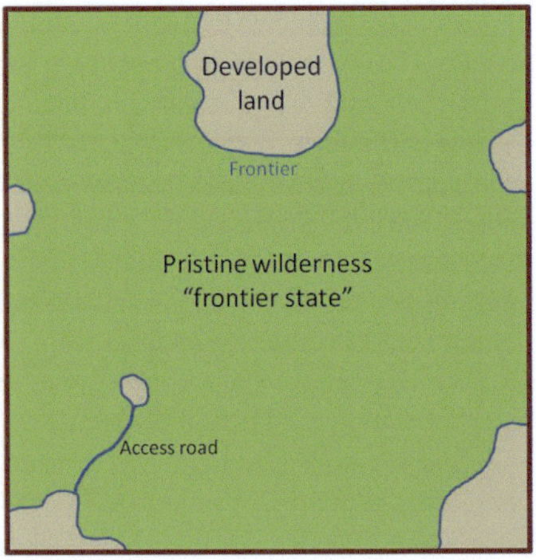

Fig. 1.1b Patterns of human development in wilderness lands: Frontier state

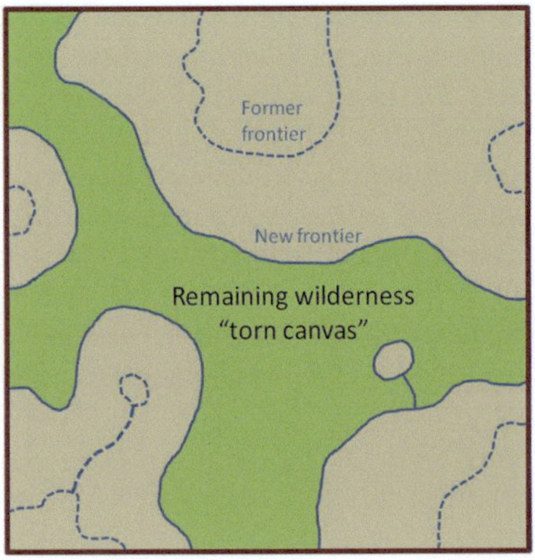

Fig. 1.1c Patterns of human development in wilderness lands: Torn canvas

Fig. 1.1d Patterns of human development in wilderness lands: Isolation

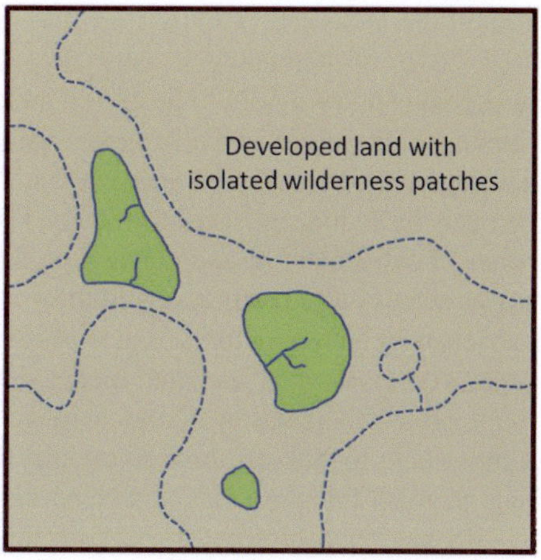

which by area is in Alaska. This pattern is similar in other large countries (e.g. Canada, Brazil, Russia and Australia) where developed land gives way to wilderness in the interior. The same can arguably be said for Europe if taken as a whole. A recent register of protected wilderness areas across the EU has revealed a pattern of isolated core areas in the central hinterland (France, Italy, Germany, Switzerland) with more extensive and connected areas in the periphery (Scandinavia and eastern Europe). Greenland, although technically part of the North American plate, is legally part of Europe (being an autonomous country within the state of Denmark) and is, by anyone's definition, mainly wilderness. As with Alaska and the USA, Greenland (together with Svalbard) greatly skews the distribution of legally protected wilderness in Europe.

1.5 Applied Mapping

In terms of total remaining wilderness by area, the figure of 34 % from McCloskey and Spalding's 1989 reconnaissance map may seem encouraging, but we need to recognise that much of this is tied up within a limited number of biomes – mainly high latitude and desert areas. Many of the biomes that are conducive to human settlement and agriculture (e.g. temperate woodland and savannah) are highly under-represented having been exploited and modified by humans for centuries. What little remains of these biomes tend to be highly fragmented and isolated leading to a need to restore and expand these areas and improve connectivity within an otherwise human modified landscape. Left alone these small fragments will most likely succumb to gradual erosion in their extent and reduction in their genetic and compositional sustainability by dint of their isolation.

This is now a well recognised problem and much work has been done on developing a more connected view of nature and wilderness conservation based largely around the so called "Cores, Corridors and Carnivores" (CCC) model (Worboys et al. 2010) or "Greenways" (see Chap. 3). The problem with the nineteenth and twentieth century model of protected areas was one of isolation. Putting a line, and sometimes literally a fence, around a natural area and keeping it "wild" by keeping development out is all well and good but in a changing world, such a model is dangerously inflexible and a risky strategy. Protecting core wilderness areas with buffer zones of extensive use and connecting them together using corridors and stepping stones across more permeable, wildlife-friendly landscapes – bridging impermeable barriers with built structures (e.g. wildlife under/overpasses across highways) where necessary – is now an accepted model. There are several examples of such connectivity projects operating across a range of spatial scales from continental (e.g. Yellowstone-to-Yukon) through country level (e.g. the Dutch EHS) to local level (e.g. Scottish Integrated Habitat Networks).

GIS has been brought to bear on this problem, using connectivity modelling techniques and toolkits available as add-ons to existing GIS software (e.g. Corridor Design). These tend to work by creating a habitat suitability model for a target species, often a keystone carnivore, and using this to identify least cost path routes or corridors between core areas where the target species is known to inhabit. Various methods exist to identify key linkages between core areas, for example using graph theory, and use these to plan the location of eco-bridges at critical pinch points and identify corridors and intermediate stepping stones for habitat restoration work and barrier removal (Pascual-Hortal and Saura 2006).

Another area of applied wilderness mapping is in targeting areas for habitat restoration or species reintroductions. This is generally referred to as "rewilding". The best locations to actively rewild (e.g. by removing human influences such as infrastructure and land use, assisted regeneration of native vegetation or reintroduction of missing native species) can be informed by careful analysis of the wilderness quality maps described here and in this book. Rewilding if done correctly and over sufficiently long time periods (e.g. 50 years or longer), can contribute new (albeit secondary) wilderness areas. Such a progression from isolated parks, through the CCC model to rewilding and new wilderness is illustrated in Figs. 1.2a, 1.2b, and 1.2c.

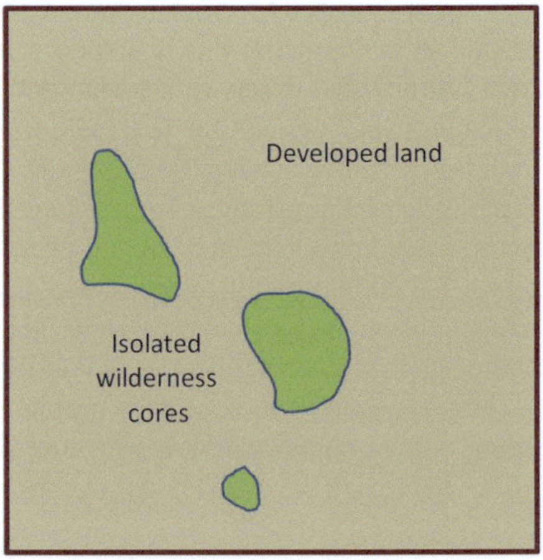

Fig. 1.2a Restoration of wilderness: Isolated core areas

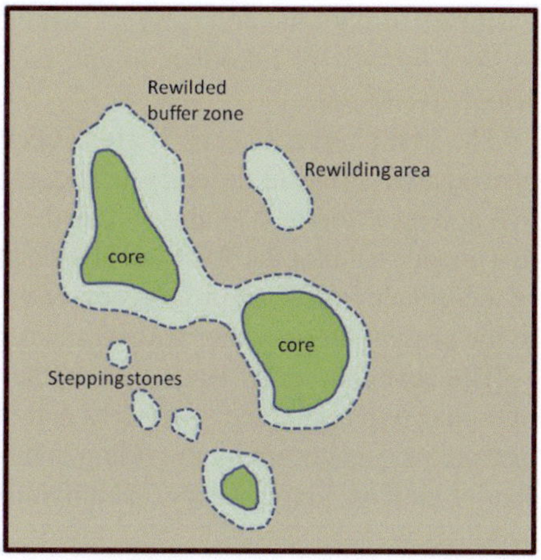

Fig. 1.2b Restoration of wilderness: Rewilding

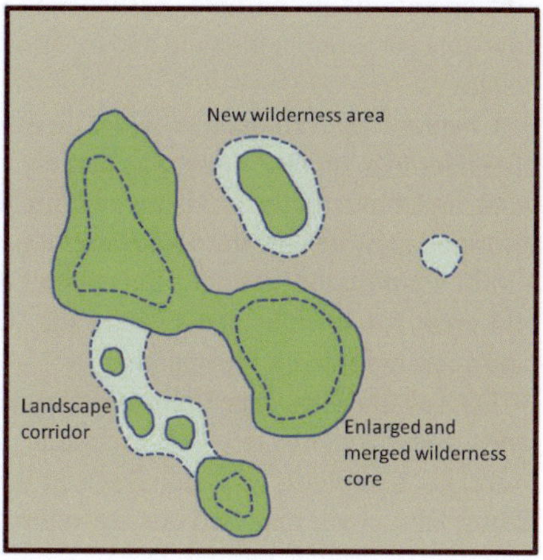

Fig. 1.2c Restoration of wilderness: Creating new wilderness

Often the simplest solution is the best. Areas that possess all the characteristics of wilderness but have none of the legal protection can be mapped against their formally protected counterparts. This is important for informing developing policies and strategies on protection as wilderness area without formal protection remain at risk from human activities and consequent degradation/reduction in wilderness quality.

1.6 The Book

The book is organised roughly into three sections, the early chapters (including this one) dealing with more conceptual and methodological approaches, the middle section dealing with certain procedural issues, and the final section providing some examples of regional and national mapping applications. These are described here in brief by way of introduction and to explain some of the logic behind chapter selection and order.

The first chapter (Chap. 2) after this is written by Rob Lesslie and gives us an historical overview of the early development of GIS-based multi-criteria approaches to mapping wilderness quality within the Australian NWI and how this was extended to a global scale for the WCMC. The chapter then describes the further development of these techniques and progression of the wilderness mapping idea in Australia up to the present day together with associated software development.

The next chapter by Roger Catchpole (Chap. 3) provides an overview of issues of connectivity and green networks and builds on the above discussion of the CCC or Greenways concept by providing detail on different mapping approaches such as spatial indices, graph theory, habitat suitability modelling, population models and agent-based models.

Chapter 4 by Lisa Machnik and colleagues looks at the practical use of spatial information technology (principally GPS and GIS) by wilderness managers to support operations in the field, particularly those concerning visitor use patterns and how data gathered in the field and inform decisions about where to allocated limited resources.

Chapter 5 by Neil Sang provides an overview of an increasingly important aspect of wilderness quality mapping, namely that of visibility analysis. Knowledge of what and how much is visible within the landscape is essential in creating an informed view of how the visitor might perceive the relative levels of human impact within a landscape setting. This chapter described various opportunities and problem areas for visibility analysis in the wilderness mapping field from a technical landscape assessment perspective.

Data availability, especially on the true levels of human impact on land cover and landscape structure, is a key potential pitfall for all wilderness quality mapping exercises. Chapter 6 by Linda See and her team develops a novel approach to validating land cover data and adding value to existing datasets through the relatively new field of crowdsourcing. Here, Google Earth imagery and "the crowd" are used to create a global map of human impact via a Geo-Wiki tool for the visualisation, crowdsourcing and validation of global land cover which can then be used to improve wilderness quality indices.

Another area of potential future development in visualisation is explored in Chap. 7 by Ben Hennig. This chapter focuses on the use of non-Cartesian geometries to display key wilderness quality variables such as remoteness as gridded cartograms. This allows remoteness to be better understood and produce high impact, thought-provoking graphs for information and visualisation purposes.

Chapter 8 by Kees Bastmeijer looks at the legal aspects of mapping wilderness and the role of law in its protection. One particular aspect of this concerns the use of GIS and other mapping approaches to inform the drawing of lines on maps to delineate and define wilderness together with the legal implications of doing so.

In Chap. 9 Mark Douglas and Bill Borrie take a long, hard philosophical look at why we may wish to map wilderness and the implications of doing so on wilderness itself. Reference is made to Heidegger's investigation of technology and links to wilderness mapping. The chapter serves as a useful "wake-up" call to wilderness mapping technologists to be careful about what it is we are mapping and why we should be mindful as to the potential for technology to remove the wildness from wilderness.

The next three chapters provide the reader with specific geographical examples of how GIS has been used to map wilderness quality. Chapter 10 by Dragos Mantoiu and colleagues is a case study of wilderness mapping for the south western Carpathian mountains in Romania. This is followed by two chapters describing national mapping programmes; Chap. 11 by Ranny Ólafsdóttir and colleagues for Iceland, and Chap. 12 by Christoph Plutzar on Austria.

Chapter 13 provides some final conclusions and closes the text with some thoughts on likely future directions and developments.

Overall, the book is designed as a reader and a marker of the current status of thinking and progress in mapping and modelling spatial patterns in wilderness quality across a range of spatial scales and for a range of applications. It is not intended to be comprehensive, rather a starting point from which one can begin to explore this fascinating and burgeoning field of endeavour within the spatial, ecological, social and cultural sciences.

References

Brun, M. NOU. (1986). GRID Arendal 1992. Source for European map: Sierra Club, World Bank, UNEP/GRID.

Carver, S., & Markieta, M. (2012). No high ground: Visualising Scotland's renewable energy landscapes using rapid viewshed assessment tools. *Proceedings of the GIS research UK 20th annual conference, 2012.* Edinburgh. http://www.geos.ed.ac.uk/~gisteac/proceedingsonline/GISRUK2012/Papers/presentation-49.pdf

Carver, S., & Wrightham, M. (2003). *Assessment of historic trends in the extent of wild land in Scotland: A pilot study.* Scottish natural heritage commissioned report no. 012 (ROAME No. F02NC11A). Edinburgh.

Carver, S. J., Comber, A., McMorran, R., & Nutter, S. (2012). A GIS model for mapping spatial patterns and distribution of wild land in Scotland'. *Landscape and Urban Planning, 104*(3–4), 395–409.

Costanza, R., d'Arge, R., de Groot, R., Farberk, S., Grasso, M., Hannon,B., Limburg, K., Naeem, S., O'Neill, R. V., Paruelo, J., Raskin, R. G., Sutton, P., & van den Belt, M. (1997). The value of the world's ecosystem services and natural capital. *Nature, 387*(6630), 253–260.

De Groot, R. S., Wilson, M. A., & Boumans, R. M. (2002). A typology for the classification, description and valuation of ecosystem functions, goods and services'. *Ecological Economics, 41*(3), 393–408.

Drewitt, A. L., & Langston, R. H. (2006). Assessing the impacts of wind farms on birds. *Ibis, 148*(s1), 29–42.

Ellis, E. C., Klein Goldewijk, K., Siebert, S., Lightman, D., & Ramankutty, N. (2010). Anthropogenic transformation of the biomes, 1700 to 2000. *Global Ecology and Biogeography, 19*(5), 589–606.

European Commission. (2013). *Guidelines on wilderness in Natura 2000: Management of terrestrial wilderness and wild areas within the Natura 2000 network.* Brussels: EU. http://ec.europa.eu/environment/nature/natura2000/wilderness/pdf/WildernessGuidelines.pdf

European Parliament Resolution of 3 February 2009 on Wilderness in Europe (Resolution 2008/2210(INI)), www.europarl.europa.eu/sides/getDoc.do?type=TA&reference=P6-TA-2009-0034&language=EN

Halpern, B. S., Walbridge, S., Selkoe, K. A., Kappel, C. V., Micheli, F., D'Agrosa, C., & Watson, R. (2008). A global map of human impact on marine ecosystems. *Science, 319*(5865), 948–952.

IUCN. (2008). *Guidelines for applying protected area management categories*. Gland: Switzerland.

Kuiters, A. T., van Eupen, M., Carver, S., Fisher, M., Kun, Z., & Vancura, V. (2013) *Wilderness register and indicator for Europe final report* (EEA Contract No: 07.0307/2011/610387/SER/B.3). Edinburgh.

Leopold, A. (1949). *A sand county almanac: With other essays on conservation from Round River*. OUP: New York.

Lesslie, R. G., & Maslen, M. (1995). *National wilderness inventory: Handbook of procedures, content and usage* (2nd ed.). Canberra: Australian Government Publishing Service.

Lesslie, R. G. (1998). *Global wilderness*. Cambridge: UNEP-WCMC. Dataset derived using the Digital Chart of the World 1993 version and methods based on the Australian National Wilderness Inventory (R. Lesslie & M. Maslen, 1995). http://www.unep-wcmc.org/resources-and-data/global-wilderness

McCloskey, M. J., & Spalding, H. (1989). A reconnaissance-level inventory of the amount of wilderness remaining in the world'. *Ambio, 18*(4), 221–227.

Nash, R. F. (1993). *Wilderness and the American mind* (3rd ed.). Yale University Press: New Haven.

Pascual-Hortal, L., & Saura, S. (2006). Comparison and development of new graph-based landscape connectivity indices: Towards the priorization of habitat patches and corridors for conservation. *Landscape Ecology, 21*(7), 959–967.

Sanderson, E. W., Jaiteh, M., Levy, M. A., Redford, K. H., Wannebo, A. V., & Woolmer G. (2002, October). The human footprint and the last of the wild. *BioScience, 52*(10), 891–904.

Schønwandt, H. K., & Dawes, P. R. (1993). An overview of Greenland's mineral exploration potential. *Rapport. Grønlands Geologiske Undersøgelse, 159*, 10–16.

Smith, J., Nayak, D. R., & Smith, P. (2014). Wind farms on undegraded peatlands are unlikely to reduce future carbon emissions. *Energy Policy, 66*, 585–591.

Scottish National Heritage. (2014). *Mapping Scotland's wildness and wild land* (Inverness: Scottish National Heritage). www.snh.gov.uk/protecting-scotlands-nature/looking-after-landscapes/landscape-policy-and-guidance/wild-land/mapping/

The Wilderness Act, Public Law 88–577 (16 U.S. C. 1131–1136) 88th Congress, Second Session 3 September 1964.

Thoreau, H. D. (1862). *Walking*, first published in 1862, Part 2, para. 18.

Tin, T., & Carver, S. (2014). A big blank white canvas? Mapping and modelling human impact in Antarctica. Science and stewardship to protect and sustain wilderness values: Tenth World Wilderness Congress symposium; October 2013; Salamanca, Spain. *Proceedings RMRS-P-64*. Fort Collins: U.S. Department of Agriculture, Forest Service, Rocky Mountain Research Station.

Warren, C. R., Lumsden, C., O'Dowd, S., & Birnie, R. V. (2005). 'Green on green': Public perceptions of wind power in Scotland and Ireland. *Journal of Environmental Planning and Management, 48*(6), 853–875.

Worboys, G., Francis, W. L., & Lockwood, M. (Eds.). (2010). *Connectivity conservation management: A global guide (with particular reference to mountain connectivity conservation)*. London: Earthscan.

Chapter 2
The Wilderness Continuum Concept and Its Application in Australia: Lessons for Modern Conservation

Rob Lesslie

Abstract Wilderness is relative; it occupies parts of a spectrum of environmental modification ranging from synthetic high-input urban and agricultural systems through to environments with minimal human interference (Lesslie RG, Taylor BG, Biol Conserv 32:309–333, 1985). This chapter considers the wilderness continuum concept which accounts for the degree to which a place is remote from and undisturbed by the influences of modern technological society, accepting that there are no absolutely inaccessible and undisturbed areas remaining on earth. The focus of the wilderness continuum concept on degrees of remoteness and naturalness in the landscape contributes to our understanding of how modern conservation landscapes can be created, including the role of larger and more intact natural areas. Discussion points to the need for comprehensive disturbance mapping and monitoring focused on patterns of land use, settlement and access across the landscape – as these represent key drivers of terrestrial environmental change. A review and discussion of Australian National Wilderness Inventory (ANWI), a wild land evaluation program conducted in Australia during the 1980s and 1990s (Lesslie RG, Maslen M, National wilderness inventory: handbook of procedures, content and usage, 2nd edn. Australian Government Publishing Service, Canberra, 1995), is provided. More recent environmental assessments that draw on the work of the ANWI are introduced. An updated global assessment of wilderness quality based on ANWI methods is presented.

Keywords Wilderness continuum • Remoteness • Naturalness • Mapping • Monitoring change • Disturbance • Modification • Fragmentation • Unmodified reference

---

R. Lesslie (✉)
Australian Bureau of Agricultural and Resource Economics and Sciences (ABARES), Canberra, Australia
e-mail: r.thackway@uq.edu.au

S.J. Carver, S. Fritz (eds.), *Mapping Wilderness*,
DOI 10.1007/978-94-017-7399-7_2

2.1 Introduction

It seems simple. The idea that wilderness epitomizes 'the big outside' – places beyond the touch of civilization where natural processes prevail (Forman and Wolke 1992). However, modern earth and conservation sciences, in highlighting the state of earth systems today, challenge that idea. Human-induced global climate change and other large-scale human interferences (including introduced species and pathogens, chemical and nutrient pollution and fire) have left no place on the surface of the earth untouched by humans. This recent change is additionally set against the recognition of widespread environmental modifications brought about by indigenous people over millennia, often through deliberate manipulation. Modern earth and conservation sciences are increasingly focused on coupled human-environment systems and the sustaining of ecosystem services. Concepts such as the 'anthropocene' are also in the ascendancy (Ellis 2011). Does wilderness have any place in twenty-first century conservation? Is the concept of wilderness simply an outmoded ideal from a bygone era?

To answer these questions we need a clear definition of wilderness – including clarity as to the attributes that define wilderness environments. To appreciate the wilderness continuum concept we also need to understand the idea that wilderness quality is a relative condition – occupying parts of a spectrum of modified environmental conditions. This spectrum ranges from synthetic high-input urban and agricultural systems at one extreme through to environments without direct human interference at the other.

This chapter argues that the rationale for the wilderness continuum concept is relevant and essential to the systematic survey and assessment of large intact natural areas, the framing of legislative and administrative approaches for their protection, and the design and implementation of modern landscape-wide nature conservation measures. The Australian National Wilderness Inventory (ANWI) (Lesslie and Maslen 1995), a wilderness mapping program completed in the late 1990s (also known as the Australian Land Disturbance Database), was based on the continuum concept. The ANWI was a fundamental exercise in natural resources inventory and assessment that continues to inform environment policy and programs in Australia and other countries.

2.2 Wilderness – Alternative Viewpoints

The notion of wilderness and its place in social-ecological discourse has evolved over the last century. This has seen concern for protecting wilderness areas extend from spiritual, aesthetic and cultural concerns in the early 1900s, to providing special recreation opportunities and habitat for iconic species to, most recently, an anchoring role in sustaining natural systems and processes. Post-war conflicts over the use of undeveloped land led to the passage of the US Wilderness Act in 1964 and the establishment of the National Wilderness Preservation System (NWPS) with mechanisms for the review and designation of wilderness on federal land. One feature of that evaluation process was the 'purism' debate, the issue being how 'wild' an area needs to be for inclusion within the NWPS.

A similar pathway was followed in Australia with measures for the protection of wilderness introduced in a number of states and nationally. By the 1990s wilderness protection had become established as a conservation objective. Wilderness, along with biodiversity and old growth forest values were, for example, explicitly identified as criteria for sustainable forest management under Australia's National Forest Policy Statement (Commonwealth of Australia 1992).

The relevance of wilderness to modern nature conservation debate is now commonly questioned (Mackey et al. 1998). It is argued, for example, that wilderness areas are unrepresentative of biodiversity and that wilderness protection often comes at the expense of the protection of more threatened or rarer habitat. A focus on wilderness is also cited as counter to the trend to landscape-wide conservation, including the promotion of off-reserve conservation management, ecological connectivity and conservation in production landscapes. More fundamentally, the concept has been criticised on the grounds that it is not measurable in any objective scientific manner – as evidenced in 'purism' debates and historically changing criteria for wilderness identification and assessment. Moreover, its cultural origin in western frontier societies is seen as difficult for societies that do not share this tradition. This particularly applies to indigenous societies where the line of separation between natural and managed landscapes is subtle or non-existent, notwithstanding the often widespread employment of powerful management tools such as fire (e.g. Gammage 2011).

The validity of these criticisms hinges very much on matters of definition and perspective. The wilderness continuum concept provides insights in this regard.

2.3 The Wilderness Continuum Concept

Differences in the definition, identification and mapping of wilderness areas prompted the suggestion by Nash (1973) and others that wilderness be considered a range of conditions at the wild end of a spectrum of remoteness and primitiveness extending from highly inaccessible and virtually undisturbed land at one end to settled land at the other. This way of viewing wilderness put a premium on variations of intensity rather than absolutes so that finding the watershed where wild becomes non-wild is made less critical.

Lesslie and Taylor (1985) took this approach a step further in introducing the wilderness continuum concept, maintaining that the attributes that characterise wilderness are remoteness and naturalness and defining wilderness quality as the extent to which a location is remote from and undisturbed by the influences of modern technological society. They argued that remoteness and naturalness are entirely relative – there being no absolutely inaccessible and undisturbed areas on Earth. It is thus possible to regard wilderness quality as existing to various degrees along the length of the continuum of remoteness and naturalness formed by the world's

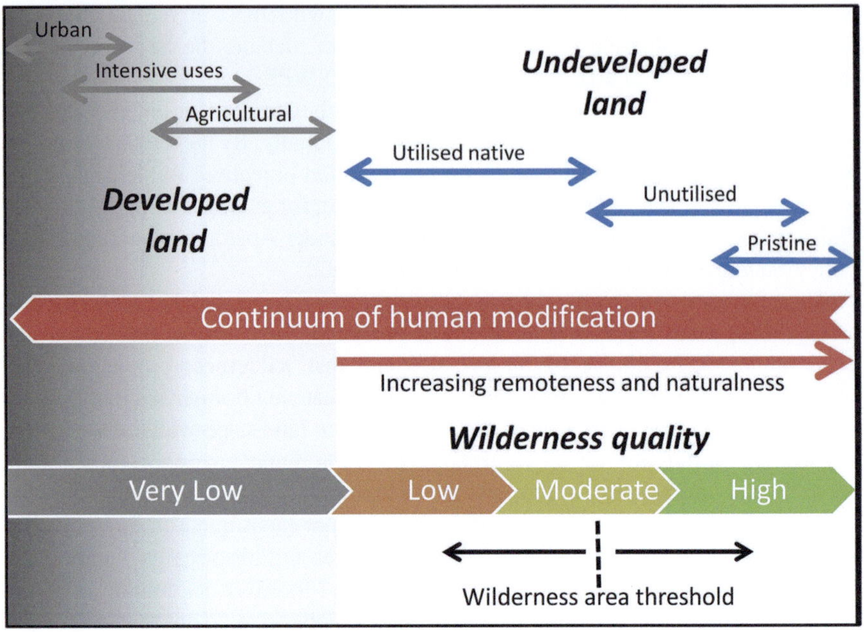

Fig. 2.1 The wilderness continuum concept (Adapted from Lesslie and Taylor 1985)

remaining areas of undeveloped land. Undeveloped land at the less remote and natural end of the continuum, such as small disturbed natural areas, can be viewed as having relatively low wilderness quality. Large intact natural areas can be viewed as having relatively higher wilderness quality. The value placed on these areas depends on context. The wilderness continuum concept is illustrated in Fig. 2.1.

The wilderness continuum concept recognises wilderness quality as a relative condition, there being no absolutely remote and undisturbed areas on Earth. The threshold or level at which remote and natural values are considered worth recognising and protecting, for example as 'wilderness areas', changes according to environmental context and over time as the demand for and supply of remote and intact land changes.

Viewing wilderness quality as a continuum of remote and natural conditions provides a solid conceptual foundation for approaching the problem of identifying wilderness resources; it also provides a coherent basis for discussion and debate regarding wilderness more broadly – from concerns about its cultural context through to measures for wilderness protection and management. A key reason for this is its focus on variability in the factors that characterise the spatial extent and impact of modern technological society in landscapes: land use and management, settlement and access.

These issues are considered further below in relation to the survey and assessment needs, the cultural context for wilderness, the protection and management of wilderness, and finally, broader lessons for conservation.

2.3.1 *Survey and Assessment*

A distinction needs to be made between (a) wilderness quality and (b) wilderness areas. Wilderness quality is the extent to which any specified unit area is remote from and undisturbed by the impacts and influence of modern technological society. Wilderness areas are relatively large intact natural areas – places where wilderness quality is defined using agreed thresholds recognized by society.

Selection criteria for wilderness areas, including thresholds of remoteness and naturalness, may be applied flexibly to single out areas having sufficient value as remote and natural places to warrant recognition. This approach accommodates the often confusing historical shifts in wilderness area identification criteria that have occurred as the supply of remote and natural land has changed and the value ascribed to these lands has evolved. Selection criteria may include factors such as size, or take into account environmental or ecological context, including the broader landscape setting. Different wilderness area selection criteria may be applied, for example, to arid or tundra environments (which are generally less developed) in contrast to temperate woodland or grassland environments (which are typically more highly affected by development and are fragmented). The application of the wilderness continuum concept to wilderness inventory and appraisal in Australia is discussed in more detail later in this chapter.

2.3.2 *Cultural Context*

By defining wilderness quality as the extent to which a location is remote from and undisturbed by the influences of modern technological society, rather than human activity *per se,* the wilderness continuum concept explicitly excludes the environmental impacts of indigenous societies. This is a critical point, noting for example in the Australian context the evidence provided by Gammage (2011) and others that that the ecology of landscapes across the continent, including forests, deserts and grasslands, have been deliberately managed and profoundly impacted by the indigenous Aboriginal population. A similar argument can perhaps be mounted for most continents with indigenous populations. Against this broader background the concept of wilderness has limited meaning. Limiting the definition to terms that describe the imprint of modern technological society distinguishes the modern definition of wilderness from earlier frontier-oriented understandings of the concept.

2.3.3 Wilderness Protection

The wilderness continuum concept offers insights into appropriate mechanisms for wilderness protection. Potentially, the wilderness quality of any area may warrant protection if, in a given context, its remoteness and naturalness is sufficiently valuable. The largest intact natural areas will generally be valued regardless of context. However, lesser degrees of remoteness and naturalness may also be recognised as important in areas that represent critical environment types (that is, the 'best of what's left'), or in other ways make significant contributions to ecological processes.

Legislation for the identification, protection and management of wilderness in the state of South Australia is based on this principle. The South Australian *Wilderness Protection Act, 1992* does not prescribe a rigid formula for the identification of wilderness areas. Wilderness criteria require that (a) the land and its ecosystems must not have been affected, or must have been affected to only a minor extent, by modern technology; and (b) the land and its ecosystems must not have been seriously affected by exotic plants or animals or other exotic organisms. Notably the wilderness quality of land may receive protection if it meets the wilderness criteria to a sufficient extent to justify its protection as wilderness or to enable it to be restored to a condition that justifies its protection as wilderness. An expert committee makes recommendations as to the potential suitability of areas for inclusion in the State's wilderness protection system. The assessment process takes account of wilderness quality measurements as well as factors such as environmental context and potential for rehabilitation.

Protected wilderness in South Australia includes relatively small intact natural areas that are now uncommon in temperate coastal regions subject to widespread development. It also includes extensive areas in the arid north of the State, including some of the best high wilderness quality locations in the world. The common feature that these protected areas share is that they represent the 'best of what's left' – the most intact examples of particular environmental settings – and for that reason alone being something worth protecting.

2.3.4 Management Principles

Application of the continuum concept to wilderness management places focus on maintaining and enhancing remoteness and naturalness. This includes protecting native species and ecological processes and controlling non-indigenous plants and animals. Regardless of the existing level of wilderness quality, management objectives remain consistent – the protection of remoteness and naturalness; ensuring areas retain or improve their remoteness and intactness. Active management may be required to ameliorate the impacts of threatening processes. Wilderness management may also provide for other uses that are compatible with the maintenance and enhancement of wilderness quality such as traditional indigenous uses, self-reliant recreation and scientific research.

2.3.5 *Lessons for Conservation*

The challenges for nature conservation in the twenty-first century include maintaining biodiversity and ecological function against a background of continuing habitat loss and fragmentation, climate change and other threatening processes including pest plants and animals and inappropriate fire regimes. Modern approaches to nature conservation aim to protect the functional integrity and resilience of natural systems in addition to more conventional approaches to protecting areas with biodiversity values. There is also an increasing emphasis on 'whole of landscape' conservation strategies and promoting connectivity.

What role can wilderness play in modern conservation? Larger, more intact natural areas have high inherent connectivity, they provide environmental benchmarks against which change can be assessed, they provide the best opportunities for effective long-term retention of species and communities and ecological processes at minimal cost (including buffering against large-scale threatening processes such as climate change and fire). In addition, these areas may be landscapes of importance to indigenous communities providing opportunities for cross-cultural and self-reliant recreation.

How large and how intact? That depends on context, such as the importance of the environmental setting to conservation (e.g. its rarity, biodiversity, endemicity, functional importance) and its level of threat. In fragmented landscapes relatively smaller-scale intact natural areas will provide corresponding conservation benefits, although to a lesser extent. Thus, both larger- and smaller-scale intact natural areas form core areas around which whole-of-landscape conservation and restoration activities can be developed, for example Australia's National Wildlife Corridors Plan (Australian Government Department of Sustainability, Environment, Water, Population and Communities 2012).

2.4 The Australian National Wilderness Inventory

The Australian National Wilderness Inventory (ANWI), an Australia-wide survey and assessment of remote and natural lands completed by the Australian Government in the late 1990s, is a prime example of the application of the wilderness continuum concept in land resources assessment (Lesslie and Maslen 1995). The ANWI developed a spatial database and analytical techniques to identify and evaluate remote and natural lands across Australia. It assisted decision-makers in delineating wilderness areas, monitoring wilderness loss, defining management options, and predicting the effect of impacts on wilderness.

The conceptual basis for the ANWI is the wilderness continuum concept. The inventory process did not take any particular biocentric or anthropocentric view of wilderness. Emphasis was placed on identifying and assessing degrees of remoteness and naturalness across the landscape using patterns of access, settlement and land use.

Four spatially-explicit wilderness quality indicators representing the two essential attributes of wilderness – remoteness and naturalness underpin the ANWI (Lesslie and Maslen 1995). The indicators are:

- ***Remoteness from settlement*** – how remote a site is from places of permanent human occupation;
- ***Remoteness from access*** – how remote a site is free from established access routes;
- ***Biophysical naturalness*** – the degree to which a site is free from biophysical disturbance caused by the influence of modern technological society;
- ***Apparent naturalness*** – the degree to which a site is free from the permanent structures associated with modern technological society.

The methods used for calculating values for these indicators reflect the constraints on available data and spatial processing capability for a continental-scale survey at that time. The type of information included land cover, land use, land management, and infrastructure.

The ANWI produced indicator values for all undeveloped areas – practically defined as areas still retaining a native vegetation cover. Values for remoteness from settlement and remoteness from access were based on calculations of distance from each survey site to the nearest settlement and access feature. A weighting regime was applied to each site; the final remoteness values of a site reflecting the greater influence of, for example, a small town compared with a single farmhouse or a highway compared with a vehicle track. Values for the apparent naturalness indicator were similarly produced, using measures of weighted distance to all structures.

The measurement of Biophysical Naturalness was approached by assuming the degree of anthropogenic change sustained in an ecosystem is directly related to the intensity and duration of land management practices associated with particular land use types. The ANWI used two rating procedures, based on five levels of land use intensity for livestock grazing and timber harvesting. The first procedure was applied to regions of the continent where arid and semi-arid livestock grazing predominates and where livestock distribution is controlled by the location of watering points. The intensity of grazing was rated according to the grazing suitability of range type, the proximity of permanent water, and tenure. The second procedure was applied to regions where grazing is less restricted by the availability of water or where commercial timber harvesting takes place. Sites were rated according to the intensity and duration of logging and grazing activity.

A total wilderness quality index was produced by combining standardised indicator values. The standard ANWI process was unweighted additive although the methods provided the ability to weight the contribution of individual indicators, and to apply criteria to account for other needs such as minimum indicators thresholds.

The resulting spatial pattern of wilderness quality assessed at a resolution of 500 m across Australia is shown in Fig. 2.2.

The map in Fig. 2.2 shows the distribution of wilderness quality across Australia based on the results of the ANWI (Lesslie and Maslen 1995). (Survey incomplete in far south-western Australia as at 1995; additional survey work completed for limited areas since 1995.) The threshold at which 'wilderness' is recognised changes according to environmental context and over time. The map shows areas of potential national significance as wilderness delineated using a set of area selection criteria and additional assessments to validate and revise ANWI results (survey not completed for Western Australia) (Australian Government Department of Sustainability, Environment, Water, Population and Communities 2008).

The continental pattern of landscape modification reflects Australia's history of exploration, settlement and development since European settlement in the late 1700s. Major urban centres and more intensive agricultural development in Australia are concentrated in the temperate regions of the east and south-east. Remoteness and naturalness values are consequently generally lower in these regions. Pastoralism and other minimal uses occur over much of the remainder. The large intact natural areas evident through central and northern Australia are arid or seasonally arid – these include some of the most extensive areas of high quality wilderness of this type in the world. These differences mean that geographic stratification is important in any analysis of results. A breakdown of these patterns by biome, for example, shows the skewed distribution of very high wilderness quality (wilderness index values greater than 18) (Table 2.1).

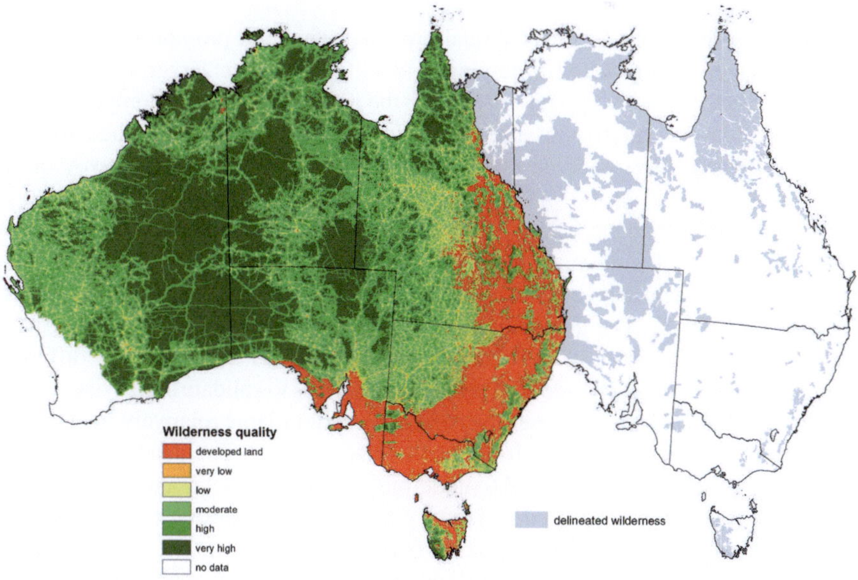

Fig. 2.2 Wilderness quality and delineated wilderness in Australia (circa 1990)

Table 2.1 Area of remaining very high wilderness quality in Australia classified using global biomes

	Very high wilderness quality	
Global biome	Area (km^2)	Proportion biome (%)
Montane Grasslands and Shrublands	0	0.0
Temperate Grasslands, Savannahs and Shrublands	2700	0.5
Temperate Broadleaf and Mixed Forests	6100	1.1
Tropical and Subtropical Moist Broadleaf Forests	400	1.3
Mediterranean Forests Woodlands and Scrubs[a]	89,100	11.1
Tropical and Subtropical Grasslands, Savannahs and Shrublands	476,200	22.3
Deserts and Xeric Shrublands	1,656,100	46.5

[a]Excludes very high wilderness quality in far south-western Australia. The area of high wilderness quality in this region is likely to be very small. Wilderness quality after ANWI; Biomes after Olsen et al. (2001). Area estimates calculated using Albers projection

More isolated locations of high wilderness quality are evident along the spine of forested ranges in eastern and south-eastern Australia. The relatively high remote and natural values in south-western Tasmania are also prominent in this context. The majority of this area is currently contained in the Tasmanian Wilderness World Heritage Area.

2.4.1 Uses of the ANWI Database

The database has been used in national processes aimed at promoting sustainable forest management in Australia, helping to identify and protect larger intact areas of native forest. Nationally agreed criteria for the establishment of a comprehensive, adequate and representative reserve system for forests in Australia include 90 % or more of high quality wilderness as measured by ANWI methods (Davey et al. 2002).

A national assessment was also completed to assist the Australian Government delineate areas of potential national significance as wilderness (Australian Government Department of Sustainability, Environment, Water, Population and Communities 2008). The delineation process involved application of a wilderness index threshold, a series of minimum size thresholds to accommodate different environmental settings and additional land assessments to validate and revise ANWI results. The result of this process, showing the extent of large relatively intact natural areas in Australia, is shown in Figure 2 (map at right). The database has been used at the national and state level to underpin wilderness assessment. The delineation process could also be used to identify core areas supporting development of landscape-scale conservation programs. One such program is the Great Eastern Ranges Initiative which aims to protect and connect intact native ecosystems along 3600 km of the Great Dividing Range and Eastern Escarpment.

2.5 Related Work

2.5.1 An International Perspective

The wilderness continuum concept and ANWI methods for survey and assessment share similarities with methods used elsewhere in the world. Surveys have been conducted in the Barents region, in parts of South America, Africa, Asia and Europe (e.g. Husby 1995). A synoptic international wilderness assessment, based on the ANWI method, has also been completed (Lesslie 1998). The assessment was based on calculations using access, settlement and infrastructure information available in the Digital Chart of the World (DCW) developed for the US Defence Mapping Agency. A more recent assessment of the distribution of global wilderness quality is shown in Figure 3; this draws on additional primary land cover and land use information in recent Globcover and global pasture datasets (Bontemps et al. 2011; Ramankutty et al. 2010). The remaining area of high wilderness quality classified by biome is shown in Table 2.2. While in the case of Australia some differences in national-level and global-level patterns are evident, spatial configurations are generally similar (comparing Figs. 2.2 and 2.3). This analysis helps place national and regional wilderness assessments in a global context. The patterns of wilderness quality in this survey also broadly reflect other remote-area global surveys produced in the last couple of decades (e.g. McCloskey and Spalding 1989; Hannah et al. 1995; Bryant et al. 1997; Sanderson et al. 2002).

Figure 2.3 is a synoptic global assessment of wilderness quality completed at 10 km resolution, based on ANWI methods. The map is an update of a global wilderness analysis completed in 1998 (Lesslie 1998) incorporating more recent land use and land cover information shown using the Robinson projection.

2.5.2 Vegetation Condition Assessment in Australia

National needs for the survey, assessment and reporting of human-induced vegetation change in Australia have benefited from the work of the ANWI and its grounding in the wilderness continuum concept. The Vegetation Assets, States and Transitions (VAST) framework has been developed as a means for ordering vegetation by degree of anthropogenic modification as a series of condition states, from a base-line condition through to total removal (Thackway and Lesslie 2008). The VAST framework accounts for change and trends in the status and condition of vegetation. The framework makes clear the links between land management and vegetation condition states and provides a mechanism for describing and tracking the resulting transitions between states caused by changes in land management practices (Thackway 2013). The VAST framework distinguishes seven condition states: Naturally bare (0), Residual (I), Modified (II), Transformed (III), Replaced (adventive) (IV), Replaced (managed) (V), and Removed (VI). The VAST

Table 2.2 Area of remaining very high wilderness quality in the world classified using global biomes

Global biome	Very high wilderness quality	
	Area (km^2)	Proportion biome (%)
Temperate grasslands, savannahs and shrublands	90,000	0.9
Mediterranean forests woodlands and scrubs	50,000	1.6
Temperate broadleaf and mixed forests	300,000	2.4
Tropical and subtropical coniferous forests	20,000	2.6
Tropical and sub tropical dry broadleaf forests	90,000	3.2
Temperate conifer forests	420,000	10.5
Mangroves	30,000	11.6
Tropical and subtropical grasslands, savannahs and shrublands	2,380,000	11.8
Flooded grasslands and savannahs	140,000	12.7
Montane grasslands and shrublands	910,000	17.6
Tropical and subtropical moist broadleaf forests	5,200,000	26.6
Deserts and xeric shrublands	8,540,000	30.7

Biomes represented in Australia

Note: Estimates not included for Tundra and Boreal Forests/Taiga due to data limitations. (Wilderness estimates modified from Lesslie 1998; biomes after Olsen et al. 2001) Area estimates calculated using Mollweide projection

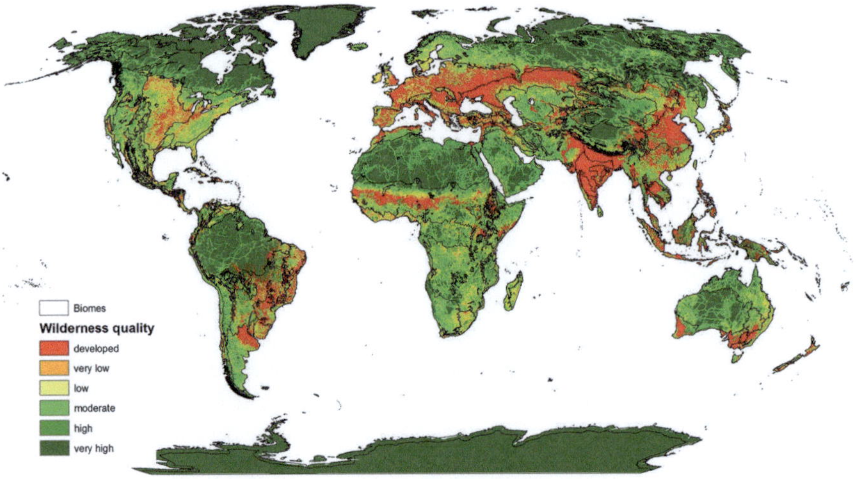

Fig. 2.3 Global wilderness assessment (Lesslie 1998)

framework prescribes a set of diagnostic criteria for each condition state including the distinction between native and non-native vegetation. The Biophysical Naturalness layer of the ANWI was a key input into an Australian VAST dataset (Lesslie et al. 2010). The VAST analysis underpins the national assessment of native vegetation condition included in the Australian report of the State of the Environment 2011 (State of the Environment 2011 Committee).

2.6 Future Directions

The perspectives and tools presented in this chapter help frame an effective approach to the survey and identification of large, relatively intact natural areas. What are current needs in advancing this capacity?

Increasing demands on land resources, including land use intensification and population growth, emphasise the need for the systematic survey and assessment of intact natural areas at local, regional and global scales. Survey information is important for these areas given the irreversible consequences of most development options. Several global surveys of human impact have been conducted, including a comprehensive global monitoring program for large intact forest landscapes (Potapov et al. 2008). However, these either target specific environments (e.g. forests) using particular size/condition thresholds (Potapov et al. 2008) or use disturbance indicators that are too generic for sufficiently precise area assessments (e.g. Sanderson et al. 2002). To provide systematic coverage at finer scales and a long-term monitoring capacity this survey work could adopt the continuum approach and be applied across all landscapes. It could also use metrics derived explicitly from settlement, infrastructure and land use features – the drivers of human-induced landscape change. Survey work of this kind has recently been completed in Europe (e.g. Fisher et al. 2010; Kuiters et al. 2013) but this requires extension elsewhere such as South America, Africa and South East Asia. Completion of further survey work, particularly in regions undergoing rapid landscape change, is a priority.

Metrics for measuring relative remoteness and intactness used for the ANWI have been improved upon in more recent European survey work (e.g. Carver et al. 2002; Fisher et al. 2010). This includes topographic and view-shed analysis for improved measurement of remoteness and apparent naturalness. There is a particular need for the development of better biophysical naturalness metrics linked to the measurement of land use and land management and broader ecological approaches to measuring the intensity and biophysical impact of human activity in landscapes (Lesslie 1997; Thackway and Lesslie 2008).

Detailed spatio-temporal data is increasingly available to support improved survey and assessment work (Stafford et al. 2012). This facilitates more accurate mapping and provides better capacity to track change. This includes tracking 'hotspots' of change and threats to remote natural lands arising from pressures such as land use intensification and climate change. Detailed survey work will also help establish priorities for the conservation of large intact natural areas as well as for investment in the complementary management of the intervening matrix, including rehabilitation. Newly available data streams include remotely sensed land cover data, digital topographic and terrain mapping and land use and dynamic vegetation.

Finally, recent advances in spatial analysis, including spatial multi-criteria analysis, enable more sophisticated contextual analysis of remote and natural lands. Factors such as biodiversity, productivity, carbon, water resources and other ecosystem services are important in considering land allocation and management priorities. The innovative spatial decision-support tool MCAS-S (ABARES 2011; Lesslie 2012), for example, is used here (Figure 4) to analyse the relationship between the spatial distribution across Australia of wilderness quality as measured by the ANWI and Net Primary Productivity as measured by mean annual net primary production (t/ha/year) from MOD17A3 data 2000–2009. The results, shown on the MCAS-S interface, point to locations where there is a coincidence of relatively high wilderness quality and high primary productivity (bottom centre map). These locations are notable along the forested ranges of eastern and south-eastern Australia, in south-western Tasmania, in Cape York Peninsula and Arnhem Land in northern Australia, and in the Great Western Woodlands of south-western Australia. These locations are a prime focus of community interest and concern for wilderness, as opposed to more extensive areas of lower productivity wilderness in the arid and semi-arid inland.

The spatial coincidence of relatively high wilderness quality and net primary productivity (bottom centre Fig. 2.4) analysed using the MCAS-S spatial decision-support tool (ABARES 2011). A matrix (far left) highlights class relationships. Total wilderness quality (as measured by the ANWI) is constructed by summing four ANWI wilderness indictors (maps at left) with equal weighting. MCAS-S features live-update functionality enabling new views to be immediately constructed

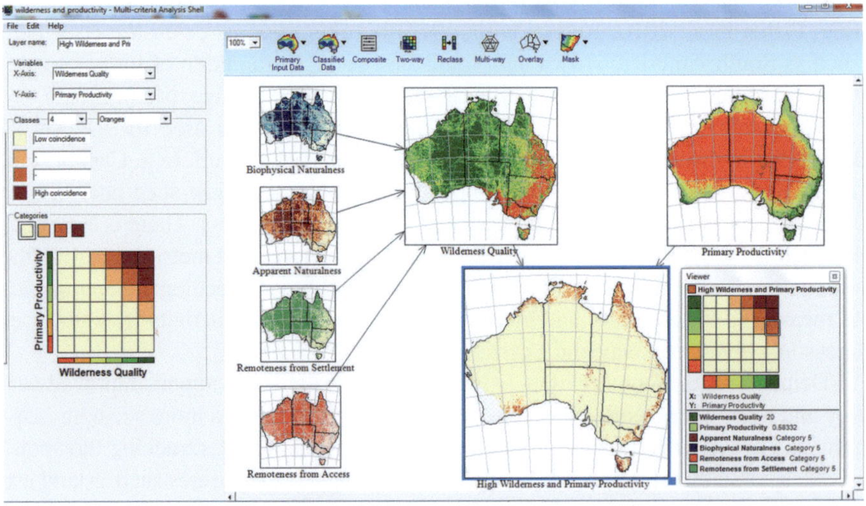

Fig. 2.4 High wilderness quality and primary productivity in Australia

and examined. The MCAS-S Viewer (bottom right) shows, at a selected point, the relationship between wilderness quality, primary productivity, and all four wilderness indicators.

Other types of spatial analyses, such as spatial connectivity analysis, also help define the role remote and natural areas can play landscape-wide conservation. Such assessments are necessary in order to place wilderness and wild-land protection in the wider framework of sustainable natural resources management and ecosystem services delivery.

2.7 Conclusion

As modern technological society extends its reach and its effects become global, those places that remain relatively remote and intact are becoming increasingly valuable. The wilderness continuum concept, in calibrating degrees of remoteness and naturalness across the landscape, contributes to our understanding of these places and options for their future management. It helps reconcile diverse interpretations of wilderness with modern views of the ecological importance of large intact natural areas, and it provides an operational basis for the identification and assessment of these resources. The ANWI is one successful example of this. Its approach has remained relevant and useful to natural resources planners and managers in Australia for over two decades, and its methods have been successfully extended internationally.

Databases like the ANWI provide the flexibility to monitor change in wilderness resources over time as land conditions change, or as previously overlooked areas become better understood and valued. These databases can also be used to examine the impact of management options or development proposals, and identify areas of potential for protection. Larger intact natural areas have in situ value as ecological reference areas supporting the continuation of evolutionary processes. They also provide the core structure for modern conservation landscapes managed for a range of ecosystem services, including biodiversity.

Spatial survey and analysis has a critical role to play in delivering on this potential. The forward agenda requires comprehensive disturbance mapping and monitoring focused on patterns of land use, settlement and access across the landscape at a range of scales from global to local. These features represent key drivers of terrestrial environmental change. New streams of satellite imagery and other spatial data describing these features mean that this basic mapping can be completed with high levels of accuracy and precision.

Once primary disturbance mapping based on land use, settlement and access is competed, next steps for spatial analysis should include:

- locating the 'the best of what's left' of key ecosystems and environments;
- identifying locations critical to the delivery of primary ecosystem services (e.g. productivity, carbon, water);

- tracking hotspots of change and threatening processes

In this way, with improved survey information, and more sophisticated analysis of the environmental, cultural and spatial context, we will be better equipped to understand the role that remaining larger and more intact natural areas can play in the future management of our landscapes, society and planet.

Acknowledgements Lucy Randall is particularly thanked for her assistance in the preparation of data for mapping. Eric Bills, Richard Thackway and Lynne Alexander, are thanked for their insightful comments on a draft which greatly contributed to the final form of the chapter.

References

ABARES. (2011). *Multi-criteria Analysis Shell for Spatial Decision Support (MCAS-S) user guide*, Australian Bureau of Agricultural and Resource Economics and Sciences, Canberra. http://www.abares.gov.au/mcass

Australian Government Department of Sustainability, Environment, Water, Population and Communities. (2012). *National Wildlife Corridors Plan: A framework for landscape-scale conservation*. Commonwealth of Australia, Canberra: National Wildlife Corridors plan Advisory Group. ISBN 978-1-921733-61-1.

Australian Government Department of Sustainability, Environment, Water, Population and Communities. (2008). *Remote and natural lands delineation*. [online]. http://www.environment.gov.au/heritage/publications/anlr/maps-delin.html. Accessed 14 Sept 2012.

Bontemps, S., Defourny, P., Van Bogaert, E., Arino, O., Kalogirou, V., & Perez, J. R. (2011). *GLOBCOVER 2009 products description and validation report*. European Space Agency, Frascati, Italy and Université Catholique de Louvain, Louvain-la-Neuve, Belgium.

Bryant, D., Nielsen, D., & Tangley, L. (1997). *The last frontier forests: Ecosystems and economies on the edge*. Washington, DC: World Resources Institute (WRI). [online]. http://www.globalforestwatch.org/english/pdfs/Last_Frontier_Forests.pdf

Carver, S., Evans, A., & Fritz, S. (2002). Wilderness attribute mapping in the United Kingdom. *International Journal of Wilderness, 8*(1), 24–29.

Commonwealth of Australia. (1992). *National forest policy statement*. Perth: Advance Press.

Davey, S. M., Hoare, J. R. L., & Rumba, K. (2002). Science and its role in Australian regional forest agreements. *International Forestry Review, 4*(1), 39–55.

Ellis, E. C. (2011). Anthropogenic transformation of the terrestrial biosphere. *Philosophical Transactions of the Royal Society, 369*, 1010–1035.

Fisher, M., Carver, S., Kun, Z., McMorran, R., Arrell, K., & Mitchell, G. (2010). *Review of status and conservation of wild land in Europe*. Project commissioned by the Scottish Government. [online]. http://www.gov.scot/Topics/Environment/Countryside/Heritage/wildland. Accessed 30 Dec 2015.

Forman, D., & Wolke, H. (1992). *The big outside. A descriptive inventory of the big wilderness areas of the United States* (Rev. ed.). Harmony Books: New York. 499 pp.

Gammage, B. (2011). *The biggest estate on earth: How Aborigines made Australia*. Crows Nest: Allen & Unwin.

Hannah, L., Carr, J. L., & Lankerani, A. (1995). Human disturbance and natural habitat: A biome level analysis of a global data set. *Biodiversity and Conservation, 4*, 128–155.

Husby, E. (1995). *Wilderness quality mapping in the Euro-Arctic Barents Region, DN-report, 1995–1994*, Trondheim: Adresseavisen Offset.

Kuiters, A. T., van Eupen, M., Carver, S., Fisher, M., Kun, Z., & Vancura, V. (2013). *Wilderness register and indicator for Europe final report* (EEA Contract No: 07.0307/2011/610387/SER/B.3). [online]. http://ec.europa.eu/environment/nature/natura2000/wilderness/pdf/Draft%20wilderness%20register.pdf

Lesslie, R. (2012). Mapping our priorities – Innovation in spatial decision support. In P. Figgis, J. Fitzsimons, & J. Irving (Eds.), *Innovation for 21st century conservation* (pp. 146–153). Sydney: Australian Committee for IUCN.

Lesslie, R. G. (1997). *A spatial analysis of human interference in terrestrial environments at landscape scale*. PhD thesis, Department of Geography, The Australian National University.

Lesslie, R. G. (1998). *Global wilderness*. Cambridge: UNEP-WCMC. Dataset derived using the Digital Chart of the World 1993 version and methods based on the Australian National Wilderness Inventory. http://www.unep-wcmc.org/resources-and-data/global-wilderness. Accessed 30 Dec 2015.

Lesslie, R. G., & Maslen, M. (1995). *National wilderness inventory: Handbook of procedures, content and usage* (2nd ed.). Canberra: Australian Government Publishing Service.

Lesslie, R. G., & Taylor, S. G. (1985). The wilderness continuum concept and its implications for Australian wilderness preservation policy. *Biological Conservation, 32*, 309–333.

Lesslie, R., Thackway, R., & Smith, J. (2010). *A national-level vegetation assets, states and transitions (VAST) dataset for Australia (version 2.0)*. Canberra: Bureau of Rural Sciences.

McCloskey, J. M., & Spalding, H. (1989). A reconnaissance level inventory of the amount of wilderness remaining in the world. *Ambio, 18*(4), 221–227.

Mackey, B. G., Lesslie, R. G., Lindenmayer, D. B., & Nix, H. A. (1998). Wilderness and its place in nature conservation in Australia. *Pacific Conservation Biology, 4*(3), 182–185.

Nash, R. F. (1973). *Wilderness and the American mind* (2nd ed.). New Haven: Yale University Press.

Olson, D. M., Dinerstein, E., Wikramanayake, E. D., Burgess, N. D., Powell, G. V. N., Underwood, E. C., D'Amico, J. A., Itoua, I., Strand, H. E., Morrison, J. C., Loucks, C. J., Allnutt, T. F., Ricketts, T. H., Kura, Y., Lamoreux, J. F., Wettengel, W. W., Hedao, P., & Kassem, K. R. (2001). Terrestrial ecoregions of the world: a new map of life on Earth. *Bioscience, 51*(11), 933–938.

Potapov, P., Yaroshenko, A., Turubanova, S., Dubinin, M., Laestadius, L., Thies, C., Aksenov, D., Egorov, A., Yesipova, Y., Glushkov, I., Karpachevskiy, M., Kostikova, A., Manisha, A., Tsybikova, E., & Zhuravleva, I. (2008). Mapping the world's intact forest landscapes by remote sensing. *Ecology and Society*, *13*(2), 51. http://www.ecologyandsociety.org/vol13/iss2/art51/

Ramankutty, N., Evan, A. T., Monfreda, C. & Foley J. A. (2010). *Global Agricultural Lands: Pastures, 2000*. Data distributed by the NASA Socioeconomic Data and Applications Center (SEDAC). http://sedac.ciesin.columbia.edu/es/aglands.html. Accessed 14 Oct 2012.

Sanderson, E. W., Jaiteh, M., Levy, M. A., Redford, K.H., Wannebo, A. V. & Woolmer, G. (2002, October). The human footprint and the last of the wild. *BioScience*, *52*(10), 891–904.

Safford, H. D., Wiens, J. A., & Uayward, G. D. (2012). The growing importance of the past in managing ecosystems of the future, pages 319–327. Chapter 24 in J. A. Wiens, G. D. Hayward, H. D. Safford, C. & Giffen (Eds.), *Historical environmental variation in conservation and natural resource management* (352 pp). Oxford, UK: Wiley-Blackwell.

State of the Environment 2011 Committee. (2011). *Australia state of the environment 2011* (Independent report to the Australian Government Minister for Sustainability, Environment, Water, Population and Communities). Canberra: DSEWPaC.

Thackway, R. (2013). Applying a system for tracking the changes in vegetation condition to Australia's forests. *Proceedings of the Institute of Foresters Australia conference, managing our forests into the 21st century, held in Canberra*, 7–11 April 2013, Canberra.

Thackway, R., & Lesslie, R. (2008). Describing and mapping human-induced vegetation change in the Australian landscape. *Environmental Management, 42*(4), 572–590.

Chapter 3
Connectivity, Networks, Cores and Corridors

Roger D.J. Catchpole

Abstract Ecological networks and connectivity assessment have a major role to play in supporting the adaptation and persistence of wildland areas as robust, functioning ecological units. They seek to achieve this by supporting the movement of species between discrete habitat patches. The 'greenways' concept developed in parallel but has a different focus that stresses the utility of linear corridors for humans. Greenways are typically defined by identifying the co-location of natural and cultural assets rather than through the dispersal of species or the location of habitat patches. When planning an ecological network it's important to explicitly consider connectivity rather than use design-led approaches that simply 'join the dots' between protected areas. This process should focus on the functional and physical linkages that are already present in order to define and build upon any residual connectivity that might remain in a landscape. Understanding the relative benefits that ecological networks bring in relation to other conservation measures that might be applied in the same area is also important. A number of different tools are available that help define connectivity. Five broad approaches have been identified: spatial indices; graph theory approaches; habitat suitability models; spatially explicit population models; and individual-based models.

Keywords Connectivity • Spatial planning • Land management • Ecological networks • Landscape genetics • Conservation • Greenways • Least cost analysis • Individual based models • Habitat suitability models • Spatial indices • Graph theory • Spatially explicit population models • Focal species • Inbreeding • Outbreeding

R.D.J. Catchpole (✉)
School of GeoSciences, University of Edinburgh, Edinburgh, UK
e-mail: roger.catchpole@gmail.com

S.J. Carver, S. Fritz (eds.), *Mapping Wilderness*,
DOI 10.1007/978-94-017-7399-7_3

3.1 Introduction

Wildlands are an ever-decreasing resource in the Anthropocene era. Since the 1800s the geophysical and chemical footprint of humanity has become a global phenomena. Our influence is now so pervasive that it rivals natural processes in terms of both scale and impact. Steffan et al. (2007) go as far as to say that we have pushed the Earth into a planetary *terra incognita* where we exist in an increasingly homogenised world with less biodiversity, fewer forests and more unstable ecosystems. Atmospheric CO^2 levels have recently exceeded 400 ppm from pre-industrial levels of 280 ppm which is a concentration that has not been seen on Earth since the Pliocene, some 3–5 million years ago (Carrington 2013). A warmer world, prone to extreme weather events, has already become evident. One fifth of the world population and more than half the population of Europe have lost naked eye visibility of the Milky Way. This conclusion, as well as the global extent of human influence, is apparent from the first world atlas of artificial night sky brightness (Cinzano et al. 2001). Given our overwhelming influence, wildlands have now become islands of naturalness in an increasingly hostile sea of human dominated 'progress'. Relatively few areas now exist where the wilderness forms the matrix and human settlements the islands. Just as the remaining pockets of biodiversity in the wider environment suffer from isolation and fragmentation, so too do wildlands which means that the patterns of connectivity within and between these areas need to be understood if they are to continue to function robustly and adapt to environmental change. The following sections will outline the conceptual basis of ecological networks and connectivity and then move on to consider some tools that can be applied to manage wildlands from this perspective.

3.2 Ecological Networks

Ecological networks have been used as a tool in European land use planning since the 1970s. Although the Dutch National Ecological Network (Jongman and Bogers 2008) is perhaps the most widely known example, many other countries have adopted this approach, not just in Europe but globally. In 2001 Bennet and Wit (2001) identified a total of 119 initiatives. The two continents that were most active in the development of ecological networks at that time were Europe (35 %) and North America (25 %).

Jongman and Pungetti (2004 p. 3) define an ecological network 'as a framework of ecological components, e.g. core areas, corridors and buffer zones, which provides the physical conditions necessary for ecosystems and populations to survive in a human dominated landscape.' This is a typical definition that fundamentally divides landscapes into three basic elements of: patch, corridor and matrix. This representation, first proposed by Forman and Godron (1981), is most often applied through the analysis of landscape structure rather than through any ecological pro-

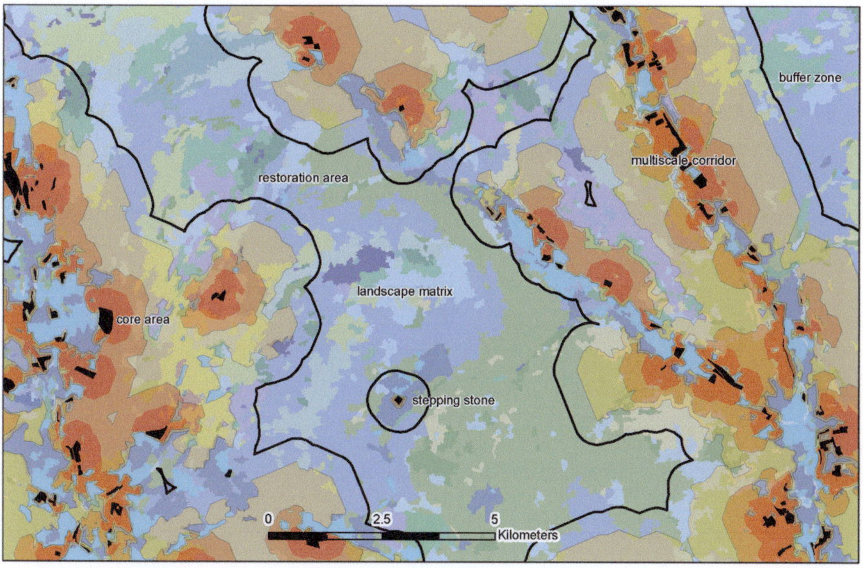

Fig. 3.1 Grassland network in the Yorkshire Dales National Park (England) calculated using a least-cost model. *Black* areas indicate existing, high biodiversity grassland patches (core areas). Semi-transparent shades of *orange* indicate an ecological network at three different scales. *Black line* indicates 500 m buffer zone around the network at the largest scale

cesses, such as dispersal. The same schematic diagram that illustrates this representation is frequently used in publications (e.g. Lawton et al. 2010) but a more helpful conceptual representation can be derived from empirical data, as shown in Fig. 3.1.

This type of representation is important as it deters non-specialists from over-simplistic interpretation of the concept which can amount to little more than 'joining the dots' when it comes to implementation (e.g. TCPA 2004). The other point that it makes is that ecological networks are formed at different scales depending on the dispersal ability of species concerned. If the focus of ecological network planning is to go beyond the management of single, charismatic focal species (e.g. Brajanoska et al. 2011) then scale is an important consideration. The continued use of other measures, such as protected areas, will remain important until a better understanding of the proportion of threatened species that will benefit from ecological networks is gained. The intensification of land management and land use change since the end of the nineteenth century has led many to conclude that protected areas cannot maintain diversity in isolation which is why ecological networks have become prominent. One of the key questions is just how much diversity can be maintained by this measure in comparison to other measures such as protected area enlargement or ecosystem service management?

The simplification of landscapes through land use change and the resulting diversity loss has been well documented (e.g. Moore 1987). This gradient of intensification can be seen in England if the current extent of priority BAP habitat (and their associated ecological networks) are considered (see Fig. 3.2).

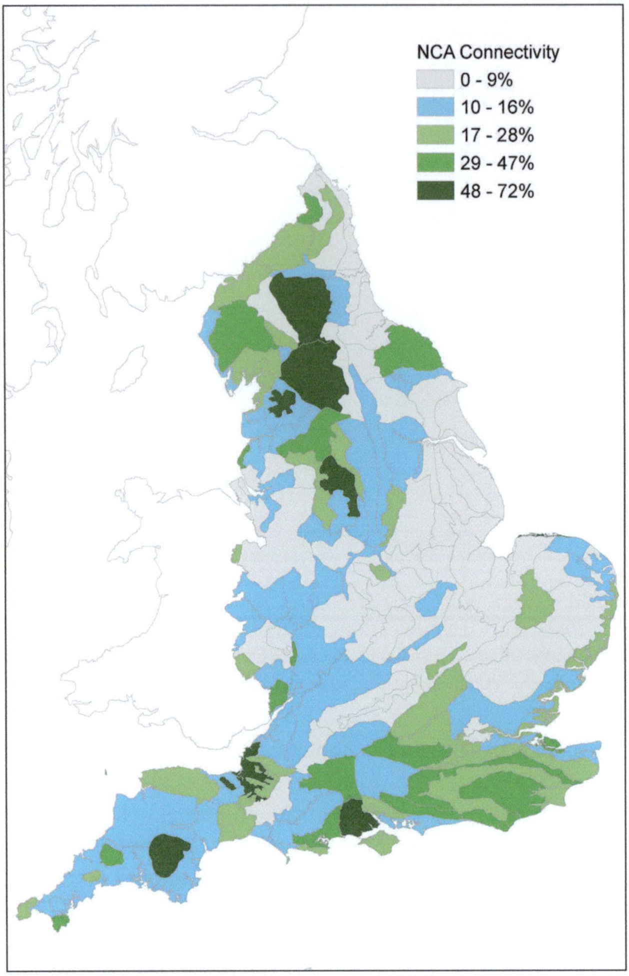

Fig. 3.2 Landscape permeability in England by biogeographical zone. Map shows ecological network coverage as a proportion of each National Character Area (NCA)

As the map demonstrates, most of the accessible fertile land in lowland areas has been converted to intensive agriculture and now only supports a depauperate assemblage of plants and animals that is, unsurprisingly, still in decline (Burns et al. 2013; Fox et al. 2013).

Although the implementation of ecological networks has been explicitly linked with spatial planning frameworks (Jongman 1995) other mechanisms, such as agri-environment schemes (if properly targeted), can also deliver this conservation measure. Jongman et al. (2004) identified three landscape ecological principles that should be incorporated into spatial planning frameworks: eco-stabilisation, connectivity and the river continuum concept.

Eco-stabilisation is a principle that was developed in the early 1980s in Central and Eastern Europe in response to the significant environmental degradation caused by Soviet era spatial planning. The key features of this concept are the designation of areas that provide environmental compensation for heavily degraded land; the linkage of these areas; and a willingness to make space available for such activities. The approach is based on the idea of a polarised landscape that was originally conceived by a Russian geographer, Rodoman, in the 1970s (Mander et al. 1995). This thinking is grounded in segregated land use planning but can be considered anachronistic from a modern, multifunctional perspective. However, it should be remembered that protected area networks also have a tendency to create polarised landscapes for biodiversity, especially where land-use in the surrounding matrix is hostile.

Connectivity is extremely important from the biological perspective and has found its expression in spatial planning through the establishment of ecological networks. In many cases ecological networks can be considered an extension of protected area network systems which means that they also have the potential to polarise landscapes, especially where significant residual connectivity remains, i.e. where patches occur in a more extensive, semi-natural landscape. Over time this approach permits intervening land-use matrix to become more hostile as the only recognised biological value (and protection) is centred upon protected areas and their associated ecological networks. Similarly where extensive land-use systems may be maintaining key linkages, abandonment will lead to succession that will then alter the nature of the matrix, potentially making it sub-optimal for the movement of key focal species. In such circumstances active management may be required rather than allowing 'rewilding' through natural succession. Connectivity issues will be elaborated further in the following section.

The River Continuum Concept was first introduced in the early 1980s by Vannote et al. (1980). It represents the first integrated perspective on the functional relationships within river catchments. It was based on observations of how macro-invertebrate dynamics and composition change along the length of river systems from the headwaters to estuaries. However, more modern perspectives view river corridors as complex ecosystems rather than just as linearly structured communities. Lateral linkages with adjacent terrestrial ecosystems, vertical linkages through riverbed substrates and temporal dynamics have all been emphasised. Flood pulse dynamics, for example, have been considered important in structuring river communities through facilitating migration, enhancing net primary production and physical restructuring of habitat (Junk et al. 1989). It has been argued that river systems should be key elements in spatial planning and the development of ecological networks because of the key role that they play in the movement of species and the wider ecosystem functions they provide. Although catchment management planning is a widely used tool, there is often a lack of integration with ecological networks (e.g. Tweed Forum 2010).

One variation on ecological networks that is subtly different, which shares more commonalities with green infrastructure, is the North American greenway concept. Ahern (2004) notes that the most widely accepted contemporary definition of greenways was given by Commission on the American Outdoors in 1987. It states that greenways provide 'people with access to open spaces close to where they live and link together the rural and urban landscapes in the American landscape, threading through cities and countryside like a giant circulation system.' A more comprehensive and inclusive definition is given by Ahern (1996 p. 132): 'Greenways are networks of land that are planned, designed and managed for multiple purposes including ecological, recreational, cultural, aesthetic, or other purposes compatible with the concept of sustainable land use.' Although the development of ecological networks and greenways has common roots, they have developed against very different political and scientific backdrops. For example, Ahern (2004) notes that one of the common arguments in favour of greenways rests on the fact that different functions and resources are typically co-located in these networks. This is in marked contrast to the protected area system in North America where national parks and wilderness areas are remote from human population centres.

Biodiversity assets, in a greenway context, occur in close proximity to where people live and work and are typically diverse in nature. A range of different features can be included in greenways, such as riparian corridors, drainage networks, derelict land, small nature reserves and linking corridors (between different networks). At larger scales, greenway corridors typically include catchment-based riparian corridors and linear upland corridors, such as whole mountain ridges. Indeed, riparian corridors are seen as fundamental to greenways because they provide connectivity, support diverse range of features as well as multiple uses and functions. This represents some common ground with ecological networks through the River Continuum Concept advocated by Jongman et al. (2004). Forman (1995) goes as far as to say that they are 'indispensable' for the sustainable functioning of any landscape because they cannot be replaced by any other feature. In other words, their contribution cannot be substituted by other means. For example, negative human impacts on the riparian zone can cause the channel to become 'disconnected' at which point the river ceases to function as a riparian corridor and only provides conduit for the species and processes that occur within the channel itself, provided good water quality can be maintained. The collateral function provided by riparian zones should not be underestimated as it includes the stabilisation of ground water flows, wildlife habitat provision, movement corridors, nutrient and sediment buffering, human recreation opportunities, well-being enhancement and support for cultural landscapes.

The co-location of cultural and natural assets has formed the basis for the definition of greenways since the early 60s (e.g. Lewis 1964). The increasing use of GIS has meant that this type of correlative, overlay analysis has since become the basis for green infrastructure planning in Europe. Indeed, it has become one of the most commonly used methods to define green infrastructure in the UK (e.g. Pankhurst 2010). However, this not only ignores functional relationships between the different factors but also the existence of ecological networks. For example, in England even though national ecological network data has been freely available since 2006 (Catchpole 2006) this information has largely been ignored by local planning authorities who are responsible for the definition and implementation of green infrastructure strategies. Although some linkages have been made with ecosystem services (e.g. Jaluzot et al. 2011) this represents another significant deficiency in integrated spatial planning. One factor that may be limiting the integration of ecological evidence could be related to the fact that only 40 % of local authorities in England directly employ ecologists and even when they are present, they are often not consulted on spatial planning issues (POST 2013). At a European level the definition of green infrastructure is far more closely aligned with ecosystem services and protected areas, rather than features such as accessible green space, as the following text indicates: 'Considering that biodiversity is also the driving force behind healthy, resilient ecosystems which, in turn, provide valuable ecosystem services, it was felt that the overall objective of an EU green infrastructure should be twofold: To safeguard and maintain Europe's biodiversity, *inter alia* by ensuring the ecological coherence of the Natura 2000 Network; and to strengthen and regenerate functional ecosystems at a broader landscape level' (DG Environment 2009 p. 4).

3.3 Connectivity

Both ecological networks and green infrastructure share a tendency towards a 'design-led' approach, especially when related to spatial planning, despite the recommendations of Jongman et al. (2004). Such approaches are strategic in nature and often lack supporting ecological evidence. Connectivity, on the other hand, is wholly concerned with ecological evidence, especially in relation to how species interact with different landscape features.

One of the first papers to explicitly define connectivity was published by Merriam (1984 p. 6) who defined 'landscape connectivity' as 'the degree to which absolute isolation is prevented by landscape elements which allow organisms to move among patches.' Taylor et al. (1993 p. 571) later refined the definition to 'the degree to which the landscape impedes or facilitate movement among resource patches.' Crooks and Sanjayan (2006 p. 4) note that 'this has since become one of the most frequently used definitions of connectivity in the scientific literature.' Both structural and functional considerations were subsequently incorporated into a single concept by With et al. (1997 p. 151) who stated that connectivity was 'the functional relationship among habitat patches, owing to the spatial contagion of habitat and the movement responses of organisms to landscape structure.' In addition to the significant insights provided by landscape ecology, metapopulation concepts have also been important in developing our understanding of connectivity (e.g. Moilenen and Hanski 2001). For those species that naturally exist within a metapopulation structure, an understanding of the process of local extinction and recolonisation is crucial if they are to be managed effectively. The persistence of a population within a geographically distinct area depends on maintaining the balance of these two processes. However, although connectivity is an important element of metapopulation studies, it is typically measured at the patch level or through nearest neighbour analysis which has been shown to provide only a very crude representation of immigration probability (Moilanen and Nieminen 2002). In order to differentiate between these different perspectives Tischendorf et al. (2003) proposed that the terms 'patch connectivity' should be only be used when talking about connectivity between a series of functionally linked patches (i.e. a metapopulation) and 'landscape connectivity' should only be used when talking about the movement between other types of patch in a wider landscape.

Ahern (2004) argues that connectivity must be understood in terms of the process or function that it's intended to support. For example, the physical connectivity required by spawning salmon is not the same as the 'functional' stepping stones required by migratory bird species. Indeed, if we consider abiotic processes, such as hydrological flows, then the definition becomes even broader. Socio-economic processes make this broader still if we think in terms of activities such as recreation, e.g. long-distance footpaths. A considerable literature is devoted to inherent value of connectivity for biodiversity, especially in the face of increasing fragmentation and land-use intensification. However, some argue that those benefits have not been sufficiently well established nor have the potential dangers outlined by Simberloff and Cox (1987) been sufficiently well explored. One of the best arguments in favour of connectivity-based conservation was put by Beier and Noss (1998 p. 1241): 'The evidence from well-designed studies suggests that corridors are valuable conservation tools. Those who would destroy the last remnants of natural connectivity should bear the burden of proving that corridor destruction will not harm target populations.' This is supported by the precautionary principle contained within the Rio Declaration on Environment and Development which states: 'In order to protect the environment, the precautionary approach shall be widely applied by states according to their capabilities. Where there are threats of serious or irreversible damage, lack of full scientific certainty shall not be used as a reason for postponing cost-effective measures to prevent environmental degradation' (UNEP 1992, Principle 15).

Confusion exists within both the scientific and conservation communities, however, about how to define connectivity and how management measures should be implemented. The idea of connectivity is more straightforward than the process of implementation. As Ahern (2004) alludes, connectivity should be considered an entirely scale- and target-dependent phenomenon. Definitions, conservation applications and measures of success all depend on the specific taxa or processes of interest as well as the spatiotemporal scales at which they occur. In spite of this confusion, two predominant perspectives have emerged from the literature in the form of structural and functional connectivity. The first is often equated with the spatial contagion of habitats and typically measured without any reference to how different organisms and processes interact with landscape features. The second also considers the distribution of habitat but includes insights into how species interact with those features at different scales. Fagan and Calabrese (2006) distinguish two types of functional connectivity with the first defining potential movement, through indirect knowledge of the species dispersal ability and the second defining actual connectivity through the quantified movement of individuals through a landscape, e.g. through radio tracking. It's important to note that structurally connected landscapes may only be functionally connected for some species and not others.

Taylor et al. (1993) stress that this distinction is not trivial and that habitats do not need to be structurally connected in order to be functionally connected. Many organisms are able to cross gaps between patches in an otherwise hostile or partially inhabitable matrix (Dale et al. 1994; Desrochers et al. 1988; Pither and Taylor 1998). Conversely, structural connectivity is not achieved if corridors fail to be utilised by the target species for which they are intended. It's generally not possible to extrapolate measurements of structural connectivity, such as inter-patch distance, and derive an overall measure of landscape connectivity because this fails to consider functional relationships such as boundary interactions, gap crossing ability and long distance dispersal. This would only be possible if the intervening matrix within a particular landscape was entirely homogenous and ecologically neutral. Obviously there are few circumstances where this might be the case. More recent derivations of incident function models do, however, include measures of functional connectivity where coefficients can be entered for differential rates of movement through different habitat types (Roland et al. 2000). Such coefficients capture the 'effective isolation' of patches (Ricketts 2001), in which movement may be reduced in some land cover types, thus showing a more realistic pattern of isolation between patches than simple Euclidean measures of distance might suggest.

The prevalence of structural connectivity measurements is largely due to the ease with which they can be calculated using a GIS. This can potentially lead to inappropriate management strategies and a diversion of attention from key issues that may be affecting population viability in a particular landscape. However, the calculation of functional connectivity is far from straightforward. As Taylor et al. (1993) note, complementarity and multiple factors determine the ability of a species to move between patches. For example, landscape context, in terms of the relative proportions of different land cover types, can determine movement behaviour for species that are ostensibly associated with a single habitat, as is the case for calopterygid damselflies in Nova Scotia. The ability of these species to cross streams and pastures to access forest patches is determined by the amount of forest in the landscape at broader spatial scales. Changes in movement behaviour were noted when clear-felling was introduced. In other words the gap-crossing ability of the species was determined by complementary habitat rather than the nature of the habitat that they were actually crossing, e.g. pasture. They also note that multiple factors can complicate matters, as was shown for flying squirrels and songbirds where movement between forest patches was a function of gap crossing ability and the successional stage of the intervening habitat. Clearly data are needed on movement rates of species through different landscape elements as well as dispersal range and boundary interaction data. Although satellite-based tracking is being used on an increasing number of species, this information will always remain limited and cannot be applied to the majority of invertebrate species that contribute most to global biodiversity.

Another factor to consider in calculating functional connectivity is that the probability of movement between patches is not symmetrical. Various landscape features within the matrix can act as either absolute or semi-permeable barriers to movement. This means that within the network patches, the probability of movement is not equal in all directions. The use of fixed buffers in geospatial analysis ignores such subtleties and assumes that species have an equal probability of movement in all directions. This type of asymmetrical landscape connectivity has been noted for Iberian lynx (Ferreras 2001). The probability of reaching another patch is also not just determined by distance. Transit times can vary considerably depending on the nature of the matrix (Schultz 1998). Movement times through more hostile areas may even be faster to avoid increased mortality, which may counteract the negative impact arising from such areas. Non-random movement can also be present which can be termed 'directional connectivity' as is the case for a cactus bug which uses olfaction to locate its prickly pear host (Schooley and Wiens 2003).

However, generalisations and the use of tacit knowledge are a necessary evil given the lack of available data. Clearly an inherent tension exists between oversimplified approaches implemented by land managers and the more ecologically robust approaches implemented by academics. Neither is satisfactory and a middle way needs to be found before landscapes become too degraded to function. In other words, we need to act on incomplete and unsatisfactory evidence, as the precautionary approach suggests. Despite the issues that have been raised, the successful integration of connectivity research with ecological network implementation is not impossible and has clearly been achieved in some instances (e.g. Watts et al. 2010). This can only occur, however, when the power of analytical tools is harnessed by ecologists who can successfully bridge the gaps between science, policy and practice. Clearly this cannot be achieved in isolation by strategic planners or GIS technicians.

3.4 Genetic Context

One of the principal drivers for the development of ecological networks and the enhancement of connectivity between habitat patches arise from the genetic consequences of isolation. Land use change over the last 60 years has led to a significant loss and fragmentation of semi-natural habitat. The consequence has been an increasing frequency of small, isolated populations. Indeed, most rare and endangered species exist in such circumstances (Holsinger and Gottlieb 1989). Fragmentation reduces the number of breeding individuals within a population and also reduces gene flow between populations (Dudash and Fenster 2000). Consequently, mating between individuals in fragmented populations is more likely to lead to inbreeding. There are a significant range of impacts. Offspring may suffer from inbreeding depression, i.e. a decline in fitness represented by poor relative performance when compared with offspring from unrelated individuals. When inbreeding persists it has been shown to cause a decline of genetic variation within a population (e.g. Schoen and Brown 1991). This can have significant negative consequences. For example, it may influence the dispersal ability and persistence of a population and lead to an increased susceptibility to pests and diseases (Barret and Kohn 1991; Frankham 1995). Furthermore, deleterious mutations occur at a higher rate and are more likely to be fixed in small populations (Lynch 1988; Lande 1994). Overall, fragmented populations may have reduced population fitness, suffer from increased extinction rates, exhibit reduced levels of genetic diversity and have higher probabilities of fixing deleterious mutations relative to more intact population structures.

Sherwin and Moritz (2000) contend, however, that the importance of genetic variation loss varies between species and is dependent upon their biology, i.e. chromosomal system, mating system and reproductive potential. It must also be remembered that some species survive with very little detectable variation at the molecular level (Reeve et al. 1990). Given the diversity of responses that are possible when genetic change occurs, it's important that conservation managers are able to use appropriate monitoring strategies to determine whether the erosion of genetic variation is actually affecting population viability or not. The authors suggest that solutions should be based on the replaceability and utility of the phenotypic variation that's expressed. This is because the historical components of genetic diversity cannot be replaced because it's accumulated over thousands of generations through random processes of drift and mutation as well as selection and adaptation (in some situations). Conservation goals should instead be focused upon key variants that determine the ability of a population to adapt to current conditions. If lost from a local population, phenotypes from elsewhere that correspond to local variants may be able to adapt relatively rapidly provided that sufficient variation is present at critical (trait-based) genetic loci. Although desirable, the identification of genetic variants that are able to adapt to future conditions is simply not possible because the future cannot be predicted. Owing to the timescale over which population genetic processes occur for longer-lived species, the majority of current genetic variation is historical and the product of past selection pressures, drift and mutation. A precautionary approach that maintains as much historical variation as possible has consequently been the main approach to preserving variation amongst conservation managers. However, the authors note that it's also important to consider variants of recent origin that may be currently more adaptive even though such variants are sometimes difficult to identify.

These contrasting perspectives can lead to different conclusions regarding the importance of connectivity and ecological networks. Whilst the widely-held assumption that increasing connectivity to 'de-fragment' landscapes is important to escape inbreeding and the loss of genetic variation, this may not be the case in all circumstances. Indeed, the introgression of genes that are not part of a local co-adapted complex may cause more harm than good when the movement of individuals is increased. This is commonly known as outbreeding depression. Outbreeding depression can occur within a single population, between geographically isolated populations or can be the product of interspecific hybridisation. It has been demonstrated by a large number of studies and occurs when a high degree of genetic distinctiveness exists between individuals which results in progeny having a lower relative fitness compared with their parents. This is either because of the dilution of genes associated with local adaptation or the disruption of co-adapted gene complexes (Fenster and Dudash 1994). Genes often interact with each other across many loci to enhance fitness, which means that such interactions can be disrupted when they are lost through the hybridisation process, even at a limited number of loci.

So how should we proceed? Much greater use needs to be made of existing knowledge as well as the insights that have been gained from landscape genetics (e.g. Holderegger et al. 2010). Significant gaps in knowledge exist for many species that should not be underestimated but much greater collaboration between science and practice, as well as between different scientific disciplines, would help land managers and spatial planners make more robust, informed judgements.

3.5 Model Selection

Once the genetic context, focus and scale of an ecological network have been chosen, the next most important judgement is the choice of spatial model. Although in most instances this will be determined by data availability, it may be possible to commission work to capture more data. Under such circumstances knowing the relative strengths and weaknesses of different approaches becomes important as simpler methods, that require a lower level of data capture, may be all that's needed. Building on Saura (2009) it's possible to identify five broad approaches: spatial indices; graph theory approaches; habitat suitability models; spatially explicit population models; and individual-based models. The amount of information required, the complexity of use and the biological realism of the models all vary between the different categories. The relative importance of these factors has been summarised in Fig. 3.3.

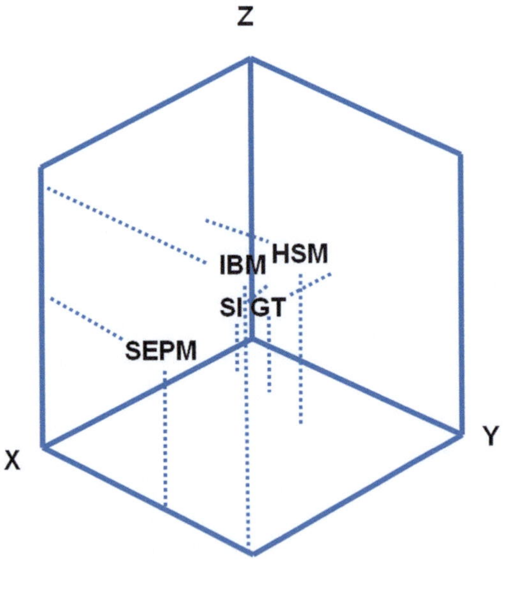

Fig. 3.3 Relative traits of ecological models that can be used to define ecological networks. *X* represents the implementation complexity of the model. *Y* represents the biological realism of the model. *Z* represents the information required to run the model which has a direct relationship with cost and practicality. *SI* spatial indices, *GT* graph theory approaches, *HSM* habitat suitability models, *SEPM* spatially explicit population models, *IBM* individual-based models

Spatial Indices (SI) are typically associated with the standard geospatial tools found in GIS packages or 'bolt-on' toolkits that calculate landscape pattern metrics, such as FRAGSTATS (McGarigal and Marks 1995). They are predominantly used to calculate structural connectivity and analyse geographical patterns. The links to any biological process, such as dispersal, are usually inferred or assumed and relatively little effort is typically made to link the observed patterns to particular processes through empirical validation. As a consequence these metrics should not be used beyond the initial characterisation and data exploration phase of a study. This lack of biological realism was well illustrated in a study by Moilanen and Nieminen (2002). They undertook a meta-analysis of a number of different studies that suggested that using nearest neighbour to define connectivity between patches was much less likely to detect statistically significant effects than studies that used more complex measures, such as incidence function models (e.g. Hanski 1994). They also compared the capacity of a number of different metrics to predict colonization events using two empirical data sets for the Glanville fritillary butterfly (*Melitaea cinxia* (L.)) and the chequered blue butterfly (*Scolitantides orion* (Pallas)). These data consisted of patch occupancy surveys from study areas in Lohja and the Åland Islands in Finland. This analysis confirmed the inferiority of nearest-neighbour measurements as well as the poor performance of buffer-based measurements where just the area and number of patches within a fixed buffer are calculated and used as a surrogate for connectivity. This was especially the case in more fragmented landscapes.

Graph Theory (GT) has been used to characterise the relationship between habitat patches through as a series of nodes and links and are calculated in a variety of different ways. Although the nodes are most often associated with habitat patches they can correspond to any landscape structure that can be identified in a geospatial analysis, such as a protected area or management unit. Although linkages are typically defined through Euclidean distance, there are more biologically realistic variants of this approach where movement can be calculated as a function of mortality risk across different land cover types (e.g. McRae et al. 2008). Euclidian distance is a simplification that assumes that species travel in straight lines between habitat patches by the shortest route. Consideration of radio tracking data reveals a quite different picture for many species and illustrates the fallacy of geographically convenient, mathematical abstractions (Hagen et al. 2011). Where perceptual scale is sufficiently large, e.g. migratory bird species, then such approaches may be defensible but it should be remembered that it is only likely to apply to a restricted range of species.

One variation of this approach that has proved popular amongst conservation biologists and resource managers is least-cost modelling (Catchpole 2006; Adriaensen et al. 2003). These are based on the assumption that different land uses have varying levels of permeability or resistance when the movement of individual species is considered. Different land cover types, derived from remote sensing imagery, are assigned permeability values that reflect the relative cost or difficulty of movement that a species is likely to encounter. A maximum dispersal distance is set but the degree of connectivity between patches is determined by the nature of the intervening matrix. The assumption is made that movement will be easier over land cover types that are most similar to the focal patch, e.g. a woodland species will move more easily through deciduous woodland land cover in comparison to intensive arable land cover. Although intuitively appealing, there are a number of issues. The first is that the movement costs are frequently derived from expert judgement rather than empirically observed data (e.g. Moseley et al. 2008). The second is that rare, long-distance dispersal events are thought to play an important role in patch colonisation (Nathan 2006) and cannot be quantified using this technique. The third is that when generic focal species approaches are used (e.g. Verbeylen et al. 2003) it is very difficult to determine the species that might be associated with the networks that have been defined.

Habitat Suitability Models (HSM) are used to characterise functional relationships between a species and a set of environmental variables that are known to have an influence on its distribution. These are what some have termed 'eco-geographical variables' (Hirzel et al. 2002). The relationship that a species has with the different variables can either be defined through deductive or inductive methods (Amici et al. 2010). The former relies on spatial data and known biological traits (e.g. Guisan and Zimmermann 2000) whilst the latter relies on field observation and measurement (e.g. Glenz et al. 2001). A grid-based, multivariate geospatial model is constructed that defines these relationships for each grid cell. This then enables users to identify areas that may be suitable for the species but for which there is currently no distribution data. They have become an increasingly important conservation tool that has been used in the development of niche-based distribution models (Hirzel et al. 2002); the definition of ecological networks (Boitani et al. 2007); the identification of species recovery areas (Cianfrani et al. 2010); and avian conservation targeting (Tirpak et al. 2009). The models are designed to utilise 'presence only' data thus avoiding the significant biases in recorder effort and ability that are typically associated with species distribution datasets. HSMs require a moderate amount of information which is usually widely available. However, the determination of the relationship between a species and the eco-geographical variables can be problematic, especially where expert judgement is used as opposed to empirical data. Another issue is that these approaches ignore biotic interactions, such as predation, that can profoundly influence species distributions (Zohmann et al. 2013). A more recent, practical application of this approach can be found in Amici et al. (2010). Instead of a single species, the authors considered the relationship between a range of environmental variables and 42 terrestrial mammal species found in the Tuscany region of Italy. A fuzzy set approach was used instead of the standard Boolean classification (i.e. 1 or 0) to associate the different land cover variables with individual cells. This allowed them to create a focal species suitability map that indicated current patterns of physical connectivity for this species group. It also accounted for the uncertainty associated with the classification of vegetation gradients through the use of fuzzy sets.

Spatially Explicit Population Models (SEPM) are based on a metapopulation approach where an assumption is made that a series of local, geographically distinct populations are connected through a dynamic process of local extinction and recolonisation. Although it is tempting to assume that species that are distributed across a series of discrete habitat patches might be interacting in this manner, validation is required before the application of such models. A much higher degree of biological realism can be found in these models owing to the fact that processes related to population viability within individual patches are directly modelled, i.e. birth, mortality, emigration and immigration. Connectivity between patches is crucial to understanding temporal variation in extinction and recolonisation but this has again, been traditionally based upon Euclidean distance. Consequently, effective isolation between patches, as determined from variation in the hostility of the landscape matrix, has generally not been considered in spite of being highlighted by authors such as Ricketts (2001). A good example of this type of model can be found in Hanski and Ovaskainen's (2000) metapopulation capacity paper. This predicts that a species is likely to persist in a landscape if the metapopulation capacity is greater than a threshold value that's determined by the life history traits of that species. This allows different landscapes to be ranked in relation to the degree to which they might support viable populations. They used a spatially realistic version of the Levins model (Levins 1969) for a finite population to model extinction and colonisation rates. This was integrated with measures of patch area, average migration distance and the distance between patches. An assumption was made that extinction rate was inversely related to patch area and that colonisation rates were determined by the proximity and size of other populations. They then demonstrated how the approach could be used to evaluate the loss or addition of habitat patches. Although still based on Euclidean distances, this type of model produces more robust, biologically realistic results provided the species biology is understood and an accurate pattern of patch occupancy can be determined.

Individual Based Models (IBM) require the greatest amount of information as they use a set of empirically observed rules that describe how individuals within a population interact with each other and their environment. Unlike population models that characterise average behaviour, the underlying assumption of IBMs is that individual behaviour is adaptive and influenced by natural selection. This will give rise to a range of strategies of varying fitness value within a natural population that are mimicked in a virtual population (Grimm and Railsback 2005). Lamberson (2012 p. 152) provides a good description that states that 'Individual-based models tend to be composed of three basic sub-models: one providing the dynamics of the environment, one providing the life history dynamics for the individuals and how they interact with other individuals and their environment, and a movement model which allows the individuals to respond to their environment in ways that may improve their fitness. Group behaviours are not imposed by these models but emerge as a result of the collective behaviour of the individuals.' As a result the analysis of this type of model 'usually involves the study of emerging patterns that result from the interactions of individuals with each other and their environment'. This is usually further refined by life history information where behaviour is modified according to life stage dependent processes such as aging. Although frequently applied to animals, such approaches have also been used to study the behaviour of plants (e.g. Campillo and Champagnat 2012). One recent study that explicitly links this approach to the definition of landscape connectivity has been undertaken by Pe'er et al. (2011). They explored the sensitivity of different components of functional connectivity using simulated landscapes with different levels of fragmentation and a hypothetical species. Using this approach they were able to explore factors such as edge interactions, different movement types (i.e. dispersal vs home range), gap crossing ability and mortality. Although this could be applied to real landscapes and species, empirical data would need to be gathered on these variables making it an unrealistic choice in most 'real world' situations.

3.6 Conclusions

Species requirements rather than cartographic convenience should be the primary consideration in the development of ecological networks. Generally wilderness areas are likely to be large and therefore capable of supporting viable populations of most species. The key task in each situation will be to identify the species for which the available area is not adequate and whether they should be managed through increasing connectivity with other wilderness areas (if possible) or through other conservation interventions, such as translocation. This process should carefully consider the relative costs and benefits between enlarging the wilderness area and increasing connectivity through the modification of the intervening landscape matrix. As Fig. 3.1 shows, such modification need not be at a large scale, just an appropriate one. Where significant residual connectivity remains, land managers and spatial planners should focus on the gaps, pinch points and barriers that remain between wilderness areas rather than on creating new 'wildlife superhighways' or adding meaningless arrows to maps that indicate desired or potential corridors. Even in England, significant residual connectivity may still be present in some areas that will be lost if they are ignored (Catchpole 2013). This can be seen at a coarser scale in Fig. 3.2. Genetic architecture should also be considered as the hybridisation of a highly adapted genotype with individuals from another population may cause the loss of genetic diversity without any apparent effect on the phenotype. Particular care needs to be taken when dealing with endemic species or populations under threat from invasive alien species. Even when larger areas of wilderness have remained intact, there may be a significant amount of internal fragmentation from roads, funicular railways, tracks, mountain bike routes and other types of trail. Again, it is important to understand which species are affected and to implement appropriate measures. When it comes to quantifying connectivity so that the context of a wilderness area can be understood and key restoration areas identified, several alternatives are possible depending on the available data. Wherever possible running more than one model and then looking for areas of coincidence or agreement may be more robust than using the outputs of a single model, as is the case in climate ensemble modelling (e.g. McSweeney et al. 2012). Generally the greatest level of biological realism should be the goal of any analysis. Where possible this should be supported through bespoke data capture although in reality this is seldom the case. This is because standard data that has been captured for other purposes, such as remotely-sensed land cover, is unlikely to have any proven functional relationship with key target species. The use of spatial indices is generally not recommended but a number of other approaches may be possible.

References

Adriaensen, F., Chardon, J. P., De Blust, G., Swinnen, E., Villalba, S., & Gulinck, H. (2003). The application of 'least-cost' modelling as a functional landscape model. *Landscape and Urban Planning, 64*, 233–247.

Ahern, J. (1996). Greenways as a planning strategy. In J. Fabos & J. Ahern (Eds.), *Greenways: The beginning of an international movement* (pp. 131–155). Amsterdam: Elsevier.

Ahern, J. (2004). Greenways in the USA: Theory, trends and prospects. In R. H. G. Jongman & G. Pungetti (Eds.), *Ecological networks and greenways: Concept, design, implementation* (pp. 34–55). Cambridge: Cambridge University Press.

Amici, V., Geri, F., & Battisti, C. (2010). An integrated method to create habitat suitability models for fragmented landscapes. *Journal for Nature Conservation, 18*(3), 215–223.

Barrett, S. C. H., & Kohn, J. (1991). Genetic and evolutionary consequences of small population size in plants: Implications for conservation. In D. A. Falk & K. E. Holsinger (Eds.), *Genetics and conservation of rare plants*. New York: Oxford University Press.

Beier, P., & Noss, R. F. (1998). Do habitat corridors provide connectivity? *Conservation Biology, 12*, 1241–1252.

Bennett, G., & Wit, P. (2001). *The development and application of ecological networks* (A review of proposals, plans and programmes). Gland: IUCN.

Boitani, L., Falcucci, A., Maiorano, L., & Montemaggiori, A. (2007). Ecological networks as conceptual frameworks or operational tools in conservation. *Conservation Biology, 21*(6), 1414–1422.

Brajanoska, R., Melovski, L., Hristovski, S., Sarov, A., & Avukatov, V. (2011). *Brown bear corridors management plan* (Report under the project: "Development of the National Ecological Network in the Republic of Macedonia (MAK-NEN)"). Skopje: Macedonian Ecological Society.

Burns, F., Eaton, M. A., Gregory, R. D., Al Fulaij, N., August, T. A., Biggs, J., Bladwell, S., Brereton, T., Brooks, D. R., Clubbe, C., Dawson, J., Dunn, E., Edwards, B., Falk, S. J., Gent, T., Gibbons, D. W., Gurney, M., Haysom, K. A., Henshaw, S., Hodgetts, N. G., Isaac, N. J. B., McLaughlin, M., Musgrove, A. J., Noble, D. G., O'Mahony, E., Pacheco, M., Roy, D. B., Sears, J., Shardlow, M., Stringer, C., Taylor, A., Thompson, P., Walker, K. J., Walton, P., Willing, M. J., Wilson, J., & Wynde, R. (2013). *State of nature report.* The State of Nature Partnership. Sandy: RSPB.

Campillo, F., & Champagnat, N. (2012). Simulation and analysis of an individual-based model for graph-structured plant dynamics. *Ecological Modelling, 234*, 93–105.

Carrington, D. (2013). *Global carbon dioxide in atmosphere passes milestone level. Climate warming greenhouse gas reaches 400 parts per million for the first time in human history*. The Guardian, Friday 10 May 2013 15.10 BST. http://www.guardian.co.uk/environment/2013/may/10/carbon-dioxide-highest-level-greenhouse-gas. Accessed 31 May 2013.

Catchpole, R. D. J. (2006). *Planning for biodiversity – Opportunity mapping and habitat networks in practice: A technical guide* (English nature research report, 687). Peterborough: English Nature.

Catchpole, R. D. J. (2013). *Ecological coherence definitions in policy and practice* (Scottish natural heritage commissioned report no. 552). Inverness: Scottish Natural Heritage; Bangkok: Food and Agriculture Organisation of the United Nations, Regional Office for Asia and the Pacific.

Cianfrani, C., Le Lay, G., Hirzel, A. H., & Loy, A. (2010). Do habitat suitability models reliably predict the recovery areas of threatened species? *Journal of Applied Ecology, 47*(2), 421–430.

Cinzano, P., Falchi, F., & Elvidge, C. D. (2001). The first World Atlas of the artificial night sky brightness. *Monthly Notes of the Royal Astronomical Society, 328*, 689–707.
Crooks, K. R., & Sanjayan, M. (2006). Connectivity conservation: Maintaining connections for nature. In K. R. Crooks & M. Sanjayan (Eds.), *Connectivity conservation* (Conservation biology, Vol. 14, pp. 1–19). Cambridge: Cambridge University Press.
Dale, V. H., Offerman, H., Pearson, S., & O'Neill, R. V. (1994). Effects of forest fragmentation on neotropical fauna. *Conservation Biology, 8*, 1027–1036.
Desrochers, A. M., Bain, J. F., & Warwick, S. I. (1988). The biology of Canadian weeds 89. *Carduus nutans* L. and *Carduus acanthoides* L. *Canadian Journal of Plant Science, 68*, 1053–1068.
Dudash, M. R., & Fenster, C. B. (2000). Inbreeding and outbreeding depression in fragmented populations. In A. G. Young & G. M. Clarke (Eds.), *Genetics, demography and viability of fragmented populations* (Conservation biology, Vol. 4, pp. 35–53). Cambridge: Cambridge University Press.
Environment, D. G. (2009). Towards a green infrastructure for Europe: Ensuring the ecological coherence of the Natura 2000 network. *European Commission Nature and Biodiversity Newsletter, 27*, 3–7.
Fagan, W. F., & Calabrese, J. M. (2006). Quantifying connectivity: Balancing metric performance with data require- ments. In K. R. Crooks & M. Sanjayan (Eds.), *Connectivity conservation* (Conservation biology, Vol. 14, pp. 297–317). Cambridge: Cambridge University Press.
Fenster, C. B., & Dudash, M. R. (1994). Genetic considerations for plant population restoration and conservation. In M. L. Bowles & C. J. Whelan (Eds.), *Restoration of endangered species: Conceptual issues, planning and implementation* (pp. 34–62). Cambridge: Cambridge University Press.
Ferreras, P. (2001). Landscape structure and asymmetrical inter-patch connectivity in a metapopulation of the endangered Iberian lynx. *Biological Conservation, 100*, 125–136.
Forman, R. T. T. (1995). *Land mosaics: The ecology of landscapes and regions*. Cambridge: Cambridge University Press.
Forman, R. T. T., & Godron, M. (1981). Patches and structural components for a landscape ecology. *Bioscience, 31*, 733–740.
Fox, R., Parsons, M. S., Chapman, J. W., Woiwod, I. P., Warren, M. S., & Brooks, D. R. (2013). *The state of Britain's larger moths 2013*. Wareham: Butterfly Conservation and Rothamsted Research.
Frankham, R. (1995). Inbreeding and extinction: A threshold effect. *Conservation Biology, 9*(4), 792–799.
Glenz, C., Massolo, A., Kuonen, D., & Schlaepfer, R. (2001). A wolf habitat suitability prediction study in Valais (Switzerland). *Landscape and Urban Planning, 55*, 55–65.
Grimm, V., & Railsback, S. F. (2005). *Individual-based modeling and ecology*. Princeton: Princeton University Press.
Guisan, A., & Zimmermann, N. E. (2000). Predictive habitat distribution models in ecology. *Ecological Modelling, 135*, 147–186.
Hagen, M., Wikelski, M., & Kissling, W. D. (2011). Space use of bumblebees (*Bombus* spp.) revealed by radiotracking. *PLoS ONE, 6*(5), e19997.
Hanski, I. (1994). A practical model of metapopulation dynamics. *Journal of Animal Ecology, 63*, 151–162.
Hanski, I., & Ovaskainen, O. (2000). The metapopulation capacity of a fragmented landscape. *Nature, 404*, 755–758.
Hirzel, A. H., Hausser, J., Chessel, D., & Perrin, N. (2002). Ecological niche factor analysis: How to compute habitat suitability maps without absence data? *Ecology, 83*, 2027–2036.
Holderegger, R., Buehler, D., Gugerli, F., & Manel, S. (2010). Landscape genetics of plants. *Trends in Plant Science, 15*(12), 675–683.
Holsinger, K. E., & Gottlieb, L. D. (1989). The conservation of rare and endangered plants. *Trends in Ecology and Evolution, 4*(7), 193–194.
Jaluzot, A., Ashton, R., Baker, R., Dean, J., Golshetti, G., Jones, N., Moss, M., Steele, M., Williams, W., & Wilmers, P. (2011). *Building natural value for sustainable economic development: The*

green infrastructure valuation toolkit user guide. London: Commission for Architecture and the Built Environment.

Jongman, R. H. G. (1995). Nature conservation planning in Europe: Developing ecological networks. *Landscape and Urban Planning, 32*(3), 169–183.

Jongman, R. H. G., & Bogers, M. (2008). *Current status of the practical implementation of ecological networks in the Netherlands* (KEN project report). Tilburg: ECNC.

Jongman, R. G. H., & Pungetti, G. (2004). Introduction: Ecological networks and greenways. In R. H. G. Jongman & G. Pungetti (Eds.), *Ecological networks and greenways: Concept, design, implementation* (pp. 1–6). Cambridge: Cambridge University Press.

Jongman, R. G. H., Külvik, M., & Kristiansen, I. (2004). European ecological networks and greenways. *Landscape and Urban Planning, 68*, 305–319.

Junk, W. J., Bayley, P. B., & Sparks, R. E. (1989). The flood pulse concept in river–floodplain systems. In D. P. Dodge (Ed.), *Canadian special publication of fisheries and aquatic sciences* (Vol. 106, pp. 110–127).

Lamberson, R. H. (2012). A brief and biased look at spatial structure in ecological models: A route to individual-based models. *Natural Resource Modeling, 25*(1), 145–167.

Lande, R. (1994). Risk of population extinction from fixation of new deleterious mutations. *Evolution, 48*(5), 1460–1469.

Lawton, J. H., Brotherton, P. N. M., Brown, V. K., Elphick, C., Fitter, A. H., Forshaw, J., Haddow, R. W., Hilborne, S., Leafe, R. N., Mace, G. M., Southgate, M. P., Sutherland, W. J., Tew, T. E., Varley, J., & Wynne, G. R. (2010). *Making space for nature: A review of England's wildlife sites and ecological network* (Report to Defra). London: DEFRA.

Levins, R. (1969). Some demographic and genetic consequences of environmental heterogeneity for biological control. *Bulletin of the Entomological Society of America, 15*, 237–240.

Lewis, P. H. (1964). Quality corridors for Wisconsin. *Landscape Architecture Quarterly*, Louisville, Kentucky, (January Issue), pp. 101–108.

Lynch, M. (1988). The rate of polygenic mutation. *Genetical Research, 51*(2), 137–148.

Mander, Ü., Palang, H., & Jagomägi, J. (1995). Ecological networks in Estonia. *Landschap, 12*(3), 27–38.

McGarigal, K., & Marks, B. J. (1995). *FRAGSTATS: Spatial pattern analysis program for quantifying landscape structure* (U.S. Forest Service general technical report PNW 351). Oregon: Oregon State University.

McRae, B. H., Dickson, B. G., Keitt, T. H., & Shah, V. B. (2008). Using circuit theory to model connectivity in ecology and conservation. *Ecology, 10*, 2712–2724.

McSweeney, C. F., Jones, R. G., & Booth, B. B. B. (2012). Selecting ensemble members to provide regional climate change information. *Journal of Climate, 25*, 7100–7121.

Merriam, G. (1984). Connectivity: A fundamental ecological characteristic of landscape pattern. In J. Brandt & P. Agger (Eds.), *Proceedings of the first international seminar on methodology in landscape ecological research and planning*. Roskilde: Roskilde University.

Moilanen, A., & Hanski, I. (2001). On the use of connectivity measures in spatial ecology. *Oikos, 95*(1), 147–151.

Moilanen, A., & Nieminen, M. (2002). Simple connectivity measures in spatial ecology. *Ecology, 83*, 1131–1145.

Moore, N. W. (1987). *The bird of time. The science and politics of nature conservation – A personal account*. Cambridge: Cambridge University Press.

Mosley, D., Ray, D., Watts, K., & Humphrey, J. (2008). *Forest habitat networks Scotland*. Contract report to Forestry Commission Scotland, Forestry Commission GB and Scottish Natural Heritage. Forest Research, Alice Holt.

Nathan, R. (2006). Long-distance dispersal of plants. *Science, 313*, 786–788.

Pankhurst, H. J. (2010). *Green infrastructure: Mainstreaming the concept. Understanding and applying the principles of green infrastructure in South Worcestershire* (Natural England commissioned reports, Number 79). Sheffield: Natural England.

Pe'er, G., Henle, K., Dislich, C., & Frank, K. (2011). Breaking functional connectivity into components: A novel approach using an individual-based model, and first outcomes. *PLoS ONE, 6*, e22355.

Pither, J., & Taylor, P. D. (1998). An experimental assessment of landscape connectivity. *Oikos, 83*, 166–174.
POST. (2013, February). *Biodiversity and planning decisions* (POSTNOTE Number 429). London: Parliamentary Office of Science and Technology.
Reeve, H. K., Westneat, D. F., Noon, W. A., Sherman, P. W., & Aquadro, C. F. (1990). DNA 'fingerprinting' reveals high levels of inbreeding in colonies of the eusocial naked mole-rat. *PNAS, 87*(7), 2496–2500.
Ricketts, T. H. (2001). The matrix matters: Effective isolation in fragmented landscapes. *The American Naturalist, 158*(1), 87–99.
Roland, J., Keyghobadi, N., & Fownes, S. (2000). Alpine *Parnassius* butterfly dispersal: Effect of landscape and population size. *Ecology, 81*, 1642–1653.
Schoen, D. J., & Brown, A. H. (1991). Intraspecific variation in population gene diversity and effective population size correlates with the mating system in plants. *PNAS, 88*(10), 4494–4497.
Schooley, R. L., & Wiens, J. A. (2003). Finding habitat patches and directional connectivity. *Oikos, 102*, 559–570.
Schultz, C. B. (1998). Dispersal behavior and its implication for reserve design in a rare Oregon butterfly. *Conservation Biology, 12*, 284–292.
Sherwin, W. B., & Moritz, C. (2000). Managing and monitoring genetic erosion. In A. G. Young & G. M. Clarke (Eds.), *Genetics, demography and viability of fragmented populations* (Conservation biology, Vol. 4, pp. 9–34). Cambridge: Cambridge University Press.
Simberloff, D., & Cox, J. (1987). Consequences and costs of conservation corridors. *Conservation Biology, 1*, 63–71.
Steffen, W., Crutzen, P. J., & McNeill, J. R. (2007). The anthropocene: Are humans now overwhelming the great forces of nature? *AMBIO: A Journal of the Human Environment, 36*(8), 614–621.
Taylor, P. D., Fahrig, L., Henein, K., & Merriam, G. (1993). Connectivity is a vital element of landscape structure. *Oikos, 68*, 571–573.
TCPA. (2004). *Biodiversity by design: A guide for sustainable communities* ('By design' guide). London: Town and Country Planning Association.
Tirpak, J. M., Jones-Farrand, D. T., Thompson, F. R., III, Twedt, D. J., Baxter, C. K., Fitzgerald, J. A., & Uihlein, W. B., III. (2009). Assessing ecoregional-scale habitat suitability index models for priority landbirds. *The Journal of Wildlife Management, 73*(8), 1307–1315.
Tischendorf, L., Darren, J., & Fahrig, L. (2003). Evaluation of patch isolation metrics in mosaic landscapes for specialist vs. generalist dispersers. *Landscape Ecology, 18*(1), 41–50.
Tweed Forum. (2010). *Tweed catchment management plan*. Roxburghshire: The Tweed Forum.
UNEP. (1992) *The Rio declaration on the environment*. http://www.unep.org/documents.multilingual/default.asp?documentid=78andarticleid=1163. Accessed 7 June 2013.
Vannote, R. L., Minshall, G. W., Cummins, K. W., Sedell, J. R., & Cushing, C. E. (1980). The river continuum concept. *Canadian Journal of Fisheries and Aquatic Science, 37*, 130–137.
Verbeylen, G., De Bruyn, L., Adriaensen, F., & Matthysen, E. (2003). Does matrix resistance influence red squirrel (*Sciurus vulgaris* L 1758) distribution in an urban landscape? *Landscape Ecology, 18*, 791–805.
Watts, K., Eycott, A. E., Handley, P., Ray, D., Humphrey, J. W., & Quine, C. P. (2010). Targeting and evaluating biodiversity conservation action within fragmented landscapes: An approach based on generic focal species and least-cost networks. *Landscape Ecology, 25*(9), 1305–1318.
With, K. A., Gardner, R. H., & Turner, M. G. (1997). Landscape connectivity and population distributions in heterogeneous environments. *Oikos, 78*, 151–169.
Zohmann, M., Pennerstorfer, J., & Nopp-Mayr, U. (2013). Modelling habitat suitability for alpine rock ptarmigan (*Lagopus muta helvetica*) combining object-based classification of IKONOS imagery and Habitat Suitability Index modelling. *Ecological Modelling, 254*, 22–32.

Chapter 4
The Use of Spatial Technology in United States Department of Agriculture Forest Service Wilderness Recreation Site Surveys

Lisa K. Machnik, Justin Ewer, and Jonathan Erickson

Abstract Understanding and managing the biophysical and social impacts of recreation in wilderness is an important component of effective stewardship. Understanding visitor use patterns can help managers determine where to focus limited resources and make informed decisions about appropriate actions. Large wilderness areas, short field seasons, and a lack of standardized protocols challenge managers to develop effective methods of data collection. Where surveys have been completed in the past they can be challenging to analyze, as protocols have varied over time and data collected on written forms may not be available in a database. As part of the Chief's 10 Year Wilderness Stewardship Challenge (10YWSC), the Pacific Northwest (PNW) Region of the U.S. Department of Agriculture Forest Service (USFS) has made great strides in completing recreation site surveys in wilderness. A contributing factor to this success was the use of spatial technology for data management and analysis. Understanding visitor use patterns can help managers determine where to focus limited resources and make informed decisions about appropriate actions.

Keywords Visitor use patterns • Impacts • Stewardship

L.K. Machnik (✉)
USDA Forest Service, Portland, OR, USA
e-mail: lmachnik@fs.fed.us

J. Ewer
Gifford Pinchot National Forest, Vancouver, WA, USA

J. Erickson
Deschutes National Forest, Bend, OR, USA

S.J. Carver, S. Fritz (eds.), *Mapping Wilderness*,
DOI 10.1007/978-94-017-7399-7_4

4.1 Introduction

Opportunities for solitude and primitive and unconfined recreation is one of the public purposes of wilderness set out in the Wilderness Act of 1964. Managers are also required to manage wilderness areas so as to leave them unimpaired for future generations (Hendee et al. 1990; Watson et al. 2000). Recreation use, however, can have impacts both on the wilderness resource as well as the experiences of other users. Collecting reliable recreation site data helps managing agencies understand and more effectively manage the biophysical and social impacts of recreation in wilderness.

In recent years, the USFS has made great strides in completing recreation site surveys in the wilderness areas it manages. A contributing factor to this success is the use of spatial technology for data management and analysis. This chapter provides examples of how managers and field staff planned for data collection and implemented field protocols; maps show how the data were analyzed and can be used as part of an effective wilderness stewardship program.

This chapter focuses on several National Forests in the PNW where wilderness recreation site inventories were conducted in 2010–2012 based on the National Minimum Recreation Site Monitoring Protocol (Cole 2006). Data collected with field GPS units was transferred to ArcGIS software (a geospatial analysis tool) for analysis. Collecting and storing data in this digital format facilitated the analysis of large amounts of data by site attribute as well as by spatial location. The use of existing commercial software allowed managers to access readily available support and ensured that wilderness managers had the ability to explore and analyze data. Collecting and maintaining data in this format helps to ensure that consistent information is available to provide an accurate frame of reference for managers to understand impacts, trends, and needs.

4.2 Background

The USFS initiated the Chief's 10 Year Wilderness Stewardship Challenge (10YWSC) as part of an effort to improve wilderness stewardship leading up to the 50th anniversary of the Wilderness Act in 2014. The 10YWSC is an outcome-based performance management and accountability system. The goal is to provide clear targets and objectives that, when achieved, will demonstrate how wilderness areas administered by the USFS are managed to a defined minimum level of stewardship (Dean 2007).

One of the ten elements of this challenge is the completion of a recreation site inventory. A completed inventory allows managers to know where and how people use each wilderness and the effects of this use. This information can be used to make decisions to protect wilderness character (for example, limits on use, revegetation of over-used areas, or education strategies).

Large wilderness areas, short field seasons, and a lack of standardized protocols challenge managers to develop effective methods of recreation site inventory data collection. Managers may not have complete data for their wilderness areas.

Wilderness areas may lack completed surveys and surveys that have been conducted in the past may be challenging to analyze, as protocols may have varied and data collected on written forms may not be available in a common database. As part of the 10YWSC, however, significant progress has been made in standardizing data and data collection processes. Guidelines for processes to follow are thorough (i.e. Cross 2010), and managers are applying new technologies at modest cost that are straightforward and relatively easy to implement.

The following examples describe how managers have implemented new protocols with commercially available equipment. It is our intent in this chapter to describe the basic requirements for planning, equipment, training, collection, and analysis so that wilderness managers can create a process useful on their unit. It is possible for managers to get useful data without a major investment in programming, infrastructure, or extensive training. Field units can use new technology to improve their knowledge of 'what's out there' and thus make better wilderness management decisions for long-term stewardship. Analysis of the data can identify areas where action is required and provides perspectives for decision-making.

4.2.1 Wilderness Ranger Corps Pacific Northwest Strike Team

During the summer of 2011, a team of interns conducted 1400 recreation site inventory surveys in thirteen wilderness areas on six National Forests in Oregon and Washington. Selecting and training interns for the task of following specific protocols during extended field assignments, ensuring adequate preparation for field time (equipment, food, first aid), and having a reliable local USFS connection were elements critical to success. Background support, including local travel, field logistics planning, and preparation for extended trips into the backcountry was managed by a Student Conservation Association (SCA) crew leader and USFS staff who served as an agency liaison while crews were in the field. Interns received initial training on site assessment protocols, modified to fit the needs of each Forest. For example, local units may have specific questions or an interest in correlating data with previously collected information.

Data collected included site attributes (e.g. size, barren core, litter, human waste, ground disturbance), photos, type of site, and locations. Using ArcMap to classify sites according to attribute data and the use of graduated symbols and colors to represent level of impact at a site provided the manager with an overview of areas of concentrated use at a glance (Fig. 4.1).

4.2.2 Pasayten and Lake Chelan-Sawtooth Wilderness

The Okanogan-Wenatchee National Forest completed a recreation site inventory across 680,000 acres of the Lake Chelan-Sawtooth Wilderness and Pasayten Wilderness in 2011–2012. Partial inventories had been completed in the past. The

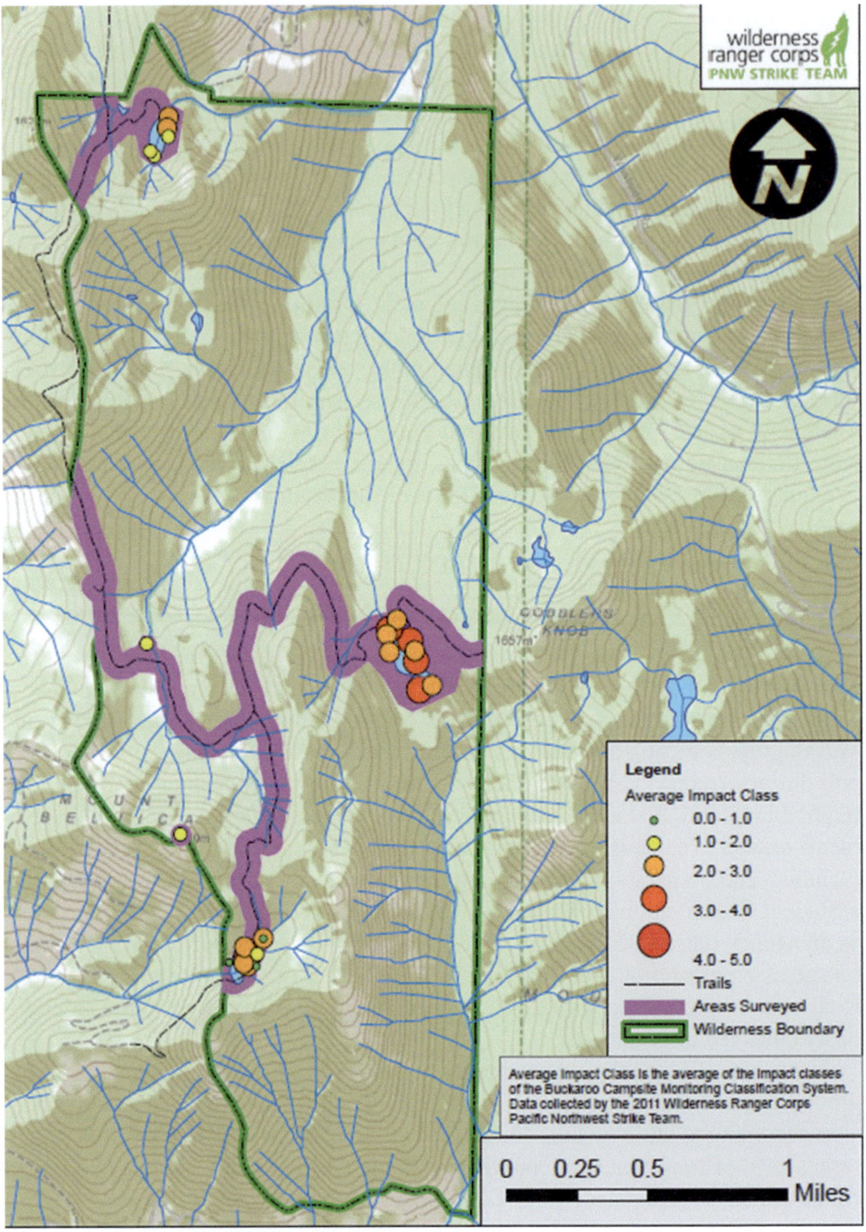

Fig. 4.1 Glacier View Wilderness campsite map

National Minimum Recreation Site Monitoring protocol was expanded to meet local planning objectives and monitoring requirements. Here, some additional attribute data were collected to provide correlation with previous surveys and Forest Plan standards to provide trend data when possible.

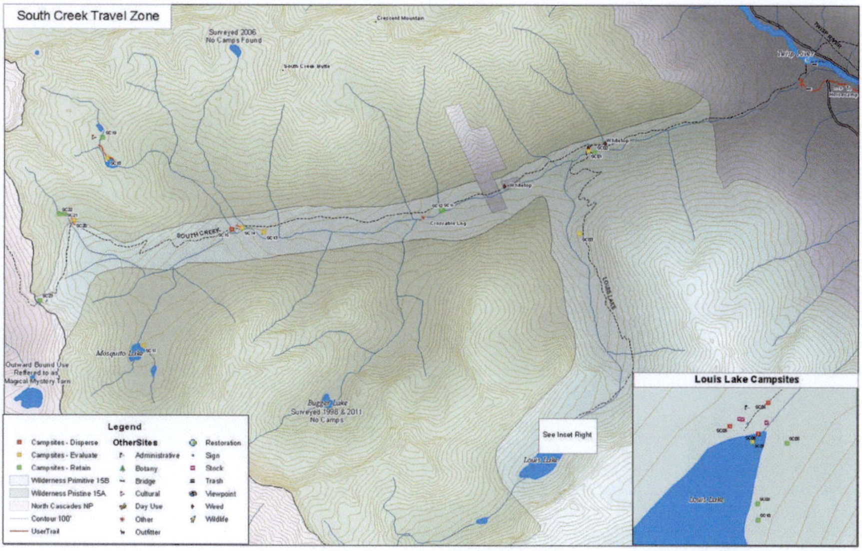

Fig. 4.2 South Creek Travel Zone map

Local expertise helped narrow the geographic areas targeted for survey to those which had a high probability of having recreation sites to narrow the work required. The project lead had previous experience with the technology to be used and had participated in similar projects in the past. Wilderness rangers, who also completed standard backcountry duties during the project, were assisted by volunteers who contributed over 1200 h of work focused on data collection. As a result of this effort:

- Site location and attributes were collected for 938 campsites;
- Data was collected for 751 'other' features, including noxious weed location, administrative features, signs, possible cultural sites, and restoration sites, and;
- Over 300 unofficial visitor-created trails were recorded.

This project also presented managers with the opportunity to gather qualitative information on non-maintained trails and other locations that do not receive regular field patrols or presence.

Figure 4.2 illustrates how the results of data collection efforts can also provide site-specific desired management direction in addition to spatial information. For example, wilderness rangers often struggle with the decision to disperse or retain campsites they encounter in the field. The site attributes collected were used to determine the level of impact at a site, and this information was incorporated into campsite symbols to indicate the desired action to be taken by a wilderness ranger. Additional information included for ranger use includes:

- Sites of administrative interest such as noxious weeds, signs, and pit toilets;
- A patrol log on the reverse side of the map, which allows wilderness rangers to reference significant details from previous patrols, and add the results of their patrol.

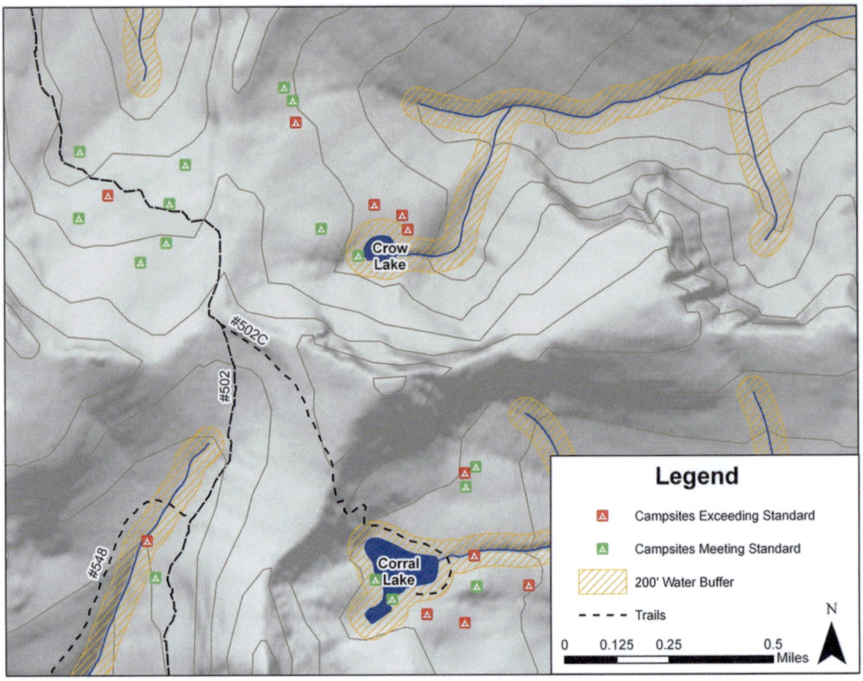

Fig. 4.3 Planning Map – Pasayten Wilderness

As can be seen in Fig. 4.3, campsites are not evenly distributed across the landscape. At a traditional map scale, popular areas such as Lewis Lake are difficult to map and read due to the high density of campsites and overlapping labels. This issue was resolved by creating an inset map. For ease of use in remote offices and field locations, this map was originally designed to be printed on an 11 × 17 sheet of paper.

The Okanogan-Wenatchee National Forest was preparing an Environmental Impact Statement (EIS), which analyzed pack and saddle stock use, and identifying which existing sites exceeded standards was an important management question. Using ArcGIS, an attribute query was run to identify which sites exceeded standards. A 200-foot buffer was also applied to waterways and lakes to help identify sites which may be of concern for fisheries. An attribute query and buffer can be completed in a few minutes by someone trained to use ArcGIS and are both tools that can be used to answer a variety of questions. Previously, correlating this information would have required a manager to review hundreds of paper records.

4.2.3 Deschutes and Willamette National Forests

In 2010, managers on the Deschutes and Willamette National Forests began project planning to conduct a census of all recreation sites in the Mt. Jefferson, Mt. Washington, Three Sisters, and Diamond Peak Wilderness Areas. Attributes that

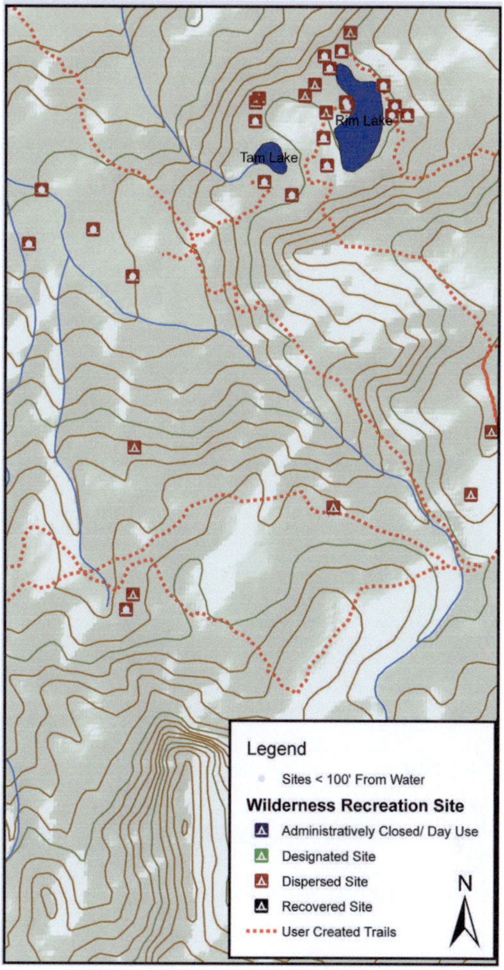

Fig. 4.4 Three Sisters Wilderness – Snow Creek Basin

could be inferred from spatial data, such as distance to water, (or were subjective) were eliminated. Next, an ArcMap personal geodatabase enabling the deployment of GIS data collection forms to multiple Trimble Juno SB GPS units was developed.

During the summer field seasons of 2011 and 2012, four USFS wilderness rangers and a volunteer crew from Lewis and Clark College collected site location, attributes, and reference photos for 3948 recreation sites, and recorded approximately 125 miles of unofficial visitor-created trails.

In Fig. 4.4, a visual representation of inventoried recreation-related impacts was overlaid with the Wilderness Resource Spectrum (WRS) prescriptions for this area. Recreation sites and user-created trails were overlaid with WRS zones. Recreation sites that are too close to water were marked with a black dot. Additional queries can easily be made to assess distance between sites, recreation site size, and which

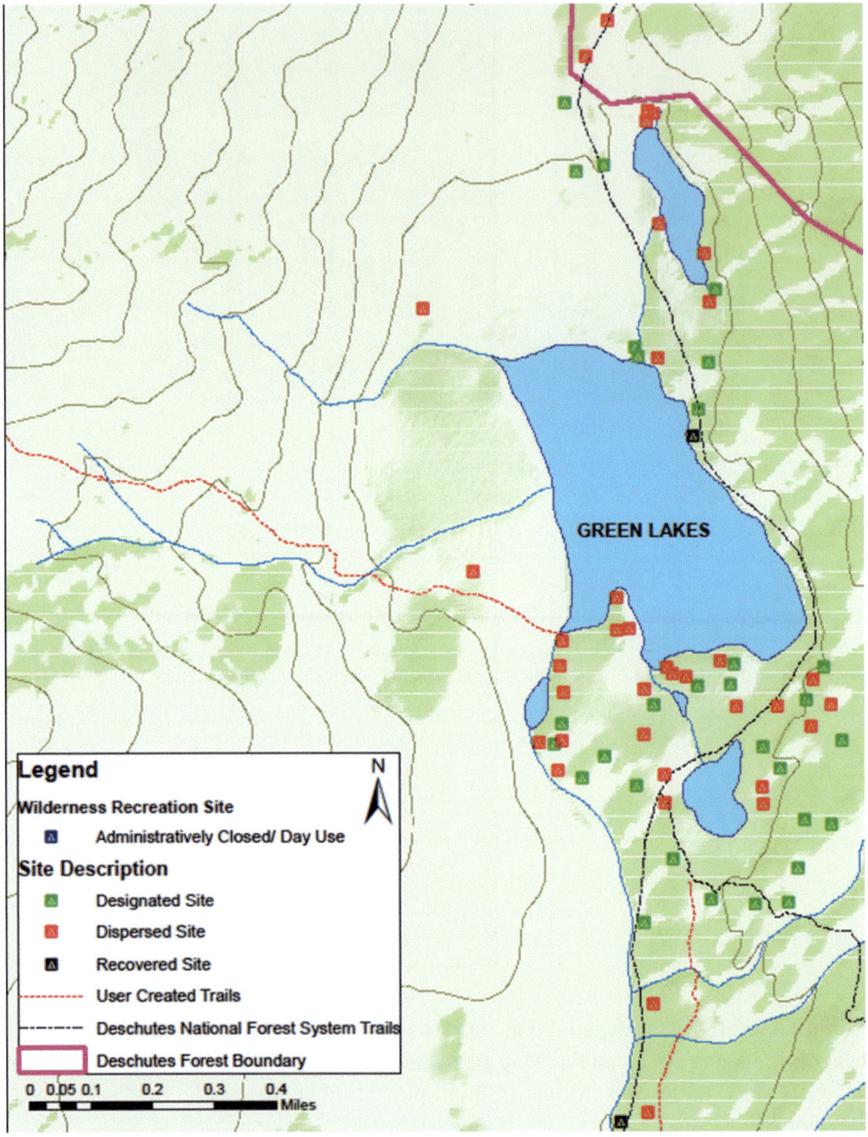

Fig. 4.5 Three Sisters Wilderness – Green Lakes Basin

sites have experienced a statistically significant change in vegetation. From this information, a detailed plan could be developed to reestablish desired WRS conditions throughout the basin.

The map shown in Fig. 4.5 was designed to help managers develop a strategic site reduction plan based on a site's description, proximity to water, proximity to other sites, and proximity to system trails. Through this process, managers can eas-

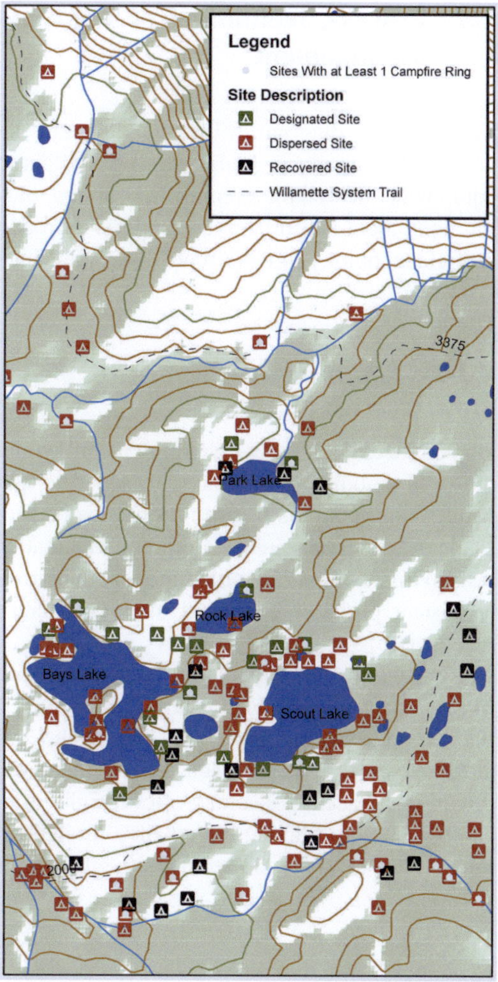

Fig. 4.6 Mt. Jefferson Wilderness – recreation sites and campfire rings

ily query site attributes to identify and map which sites are most suitable for decommission and or restoration.

Jefferson Park (Fig. 4.6) is a high use area within Mt. Jefferson Wilderness on the Willamette National Forest. Visitors are required to use designated sites. Over time, the use of the designated site system as a means to limit recreation impacts has yielded mixed results; an intense network of dispersed sites has developed among designated sites. Managers chose to identify which sites were known to have campfire rings as a starting point for a dispersed campsite restoration plan. The map in Fig. 4.6 uses typical site description icons and a black dot identifies sites with campfire rings. Additional queries can be made to identify second priority restoration activities using attributes such as proximity to system trails and water, or other parameters such as site size or change in vegetation.

4.3 Process, Technological Needs and Effective Data Management

The workflow diagram in Fig. 4.7 illustrates the general process followed in each project. Management goals varied by wilderness area, but similar protocols and equipment allowed for streamlined training that transferred across units. Using similar equipment and processes supports a broader understanding and application of spatial technology in wildereness management.

4.3.1 Planning the Project

One of the challenges to collecting useful wilderness recreation data is determining the right amount of information needed to address the management questions at hand. With hundreds or even thousands of sites within or across a complex landscape of wilderness areas, completing recreation site surveys can be a monumental task. While using these tools for data collection can speed the process in the field, a consistent protocol is critical to having data that is useful and meaningful for managers.

Reviewing thc applicable Land Management Plan (LMP) or Wilderness Management Plan (WMP) is an appropriate starting place. It can be helpful to outline the management approach (such as the WRS) and then identify the individual attributes prescribed for each area of interest. For wildernesses without planning direction, adjacent areas may offer insight, or one of the standards currently in use can be adopted. Manual counts on a limited number of pre-identified attributes of interest, for example, collecting site attribute data in a limited area to assess what

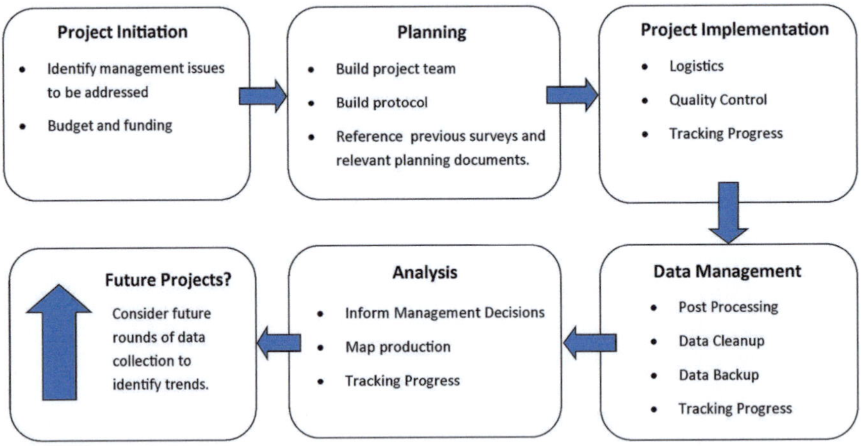

Fig. 4.7 Workflow process

types of data are needed, may be beneficial when creating management standards, and can help calibrate the future development of recreation site protocols. Value appropriate to the resource, such as camping set back requirements or campfire regulations, can be used in developing a meaningful attribute list. Another approach is to use the applicable laws and regulations that govern recreation use on the Wilderness in question.

4.3.2 Equipment

A variety of data collection equipment is readily available from commercial vendors at a range of price points. Within a government agency, other resource areas (i.e. forestry, range, wildlife) may already have, or have access to, this equipment. In the examples presented, field-going teams used a combination of Trimble Juno's, an external antenna, and laptop/desktop computers. ArcGIS software (ArcPad and ArcMap) were used to collect and view the data. Additional software used in these examples included ArcMap 10 w/Xtools, S1 Data Mgr, and Microsoft Excel.

Testing equipment in a variety of field conditions prior to committing to its use will help ensure that protocols and processes function as intended and that data is complete and reliable. In these examples batteries typically lasted 1–1.5 days in the field, with use of approximately 0.75 days during that time. Field users experimented with solar charging but were not able to eliminate battery use.

At the conclusion of each data collection period either the Lead Wilderness Ranger or crew leader would "check-in" recreation site data from each collection unit and then review and verify the data within the ArcMap attribute table. Using the optional XTools within ArcMap, data are immediately available and can be exported into a Microsoft Excel table. While export into Excel is not necessary to work with the data, this allows data analysis on demand and provided immediate feedback to users. Finally, all associated data, map products, and photos were backed up to the U.S. Forest Service corporate network drives and onto an external hard drive.

4.3.3 Data Management

Protocols for access to data and file management should be predetermined and consistent. The following steps can reduce the risk of data loss or corruption:

- Determine who should have access to the data during the collection process. Assigning data management to one individual may reduce risks to the integrity of the information, where there is danger of the collector(s) inadvertently deleting or editing source data. One control option is to leave the check-in/check-out process to a project lead or manager rather than field data collectors.

- Set up a file structure and naming protocol beforehand. Developing or changing this during the process requires additional time and increases the possibility of inadvertently deleting data.
- ArcPad's ability to check out a geodatabase for disconnected editing on a handheld device eliminates the need for data entry, or managing multiple databases. A potential risk with this process is for two field collectors to collect new data or edit the same site. This risk can be mitigated by sound field logistics and by checking collected data for inaccuracies or duplicate records.

The USFS uses network drives and cloud computing (Citrix). One advantage is that users can access the same information from multiple locations. A disadvantage is that a portion of this platform has been designed internally and does not support the ArcPad check-in/check-out feature. While work is underway to support this functionality, this experience serves as a reminder to ensure that systems are compatible when introducing new equipment or software components.

4.4 Discussion

GIS-based recreation site data provides both a visual representation and a platform for statistical analysis. The density and distribution of recreation sites across the landscape, as well as their specific attributes, can be enhanced visually to communicate components of the underlying data set. Additionally, site data can be overlaid with other data sets to analyze potential relationships between recreation and other landscape functions. For example, a cursory statistical analysis of recreation site data was used in concert with Wilderness permit data, daily Wilderness Ranger Reports, Visitor Contact reports and invasive species monitoring to derive a more holistic picture of recreation use dynamics and environmental impacts. The resulting information added context to current projects related to outfitter guiding, trails management and directing the objectives of field staff. Recreation site data will play a crucial role in identifying strategic restoration projects and future recreation management prescriptions.

Planning for data display should guide the production of any maps. Managers should identify whether a map will be at a landscape or wilderness-scale versus a smaller scale like a lake basin or travel corridor. Printing and other display capacity will influence how the data is displayed as well as the usefulness of graduated symbols, labels, and reference data.

The examples presented here benefited from the use of a national minimum protocol; this protocol allowed field staff to complete surveys that, while not highly sensitive at a site-specific scale, provided information about conditions across the wilderness. This initial broad picture can then be used to identify sites where a higher level of detail is needed. For example, in the Pasayten and Lake Chelan-Sawtooth Wildernesses, the critical standards were campsite area, tree damage, and site-specific indicators; campsite density was not a driving standard. Understanding visitor use patterns, such as the relative density of campsites, or expansions in campsite area can indicate where managers need to focus education and resource management efforts.

New rangers must quickly develop knowledge of the local wilderness resource. Most rangers learn over a period of several years, often rediscovering sites that prior rangers had visited but not necessarily documented. Ranger maps can cut down on the learning curve by providing 'travel zone' information that includes campsites, popular access points and points of interest for visitors, and noxious weed sites. Linking this map information to patrol logs can provide an effective method to identify and track issues over time.

In the examples provided, the ease of data collection in a digital format saved significant time that would otherwise be required for data entry. For example, after a single field visit, identifying multiple sites within 100 ft of water can be completed in minutes and mapped with other features for analysis, planning, and monitoring needs. Traditional 'pacing and paper' data collection adds a layer of time and complexity that can now be directed to other activities while in the field, such as visitor contacts or data collection at additional sites. In addition, the potential for field data errors was reduced and data can be exported to other corporate database formats. Using this technology gives managers tools to more fully realize the potential of monitoring work.

Monitoring recreation sites, including campsites, day use destinations, and the user-created trails associated with visitors seeking solitude and/or a primitive recreation experience can help track some of the physical impacts to wilderness such as proliferation and spread of campsites. Understanding visitor use patterns can help managers determine where to focus limited resources and make informed decisions about appropriate actions. The examples presented in this chapter demonstrate the use of spatial technology in wilderness management that has supported the USFS in the PNW in achieving part of the 10YWSC. Finally, this information can help managers inform visitors how to best realize not just the type of experience they seek but also how they can help minimize impacts to the resource and ensure that these wilderness areas will remain accessible and unimpaired for future generations to enjoy.

References

Cole, D. (2006). *Minimum recreation site monitoring protocol*. Recreation site monitoring procedures for element 6 of the chief's 10 year wilderness stewardship challenge. http://www.wilderness.net/index.cfm?fuse=toolboxes&sec=recsitemonitor

Cross, T. (2010). *A wilderness manager's desktop guide to recreation site inventories*. http://www.wilderness.net/toolboxes/documents/recsitemonitor/FS_10YWSC_Element_6_Guidebook.pdf

Dean, L. C. (2007). Tracking progress: Applying the forest service 10 year wilderness stewardship challenge as a model of performance management. In: Watson, A., Sproull, J., & Dean, L.C., comps. 2007. *Science and stewardship to protect and sustain wilderness values: Eighth world wilderness congress symposium: September 30 – October 6, 2005*; Anchorage, AK. Proceedings

RMRS-P-49. Fort Collins, CO: U.S. Department of Agriculture, Forest Service, Rocky Mountain Research Station.

Hendee, J. C., Stankey, G. H., & Lucas, R. C. (1990). *Wilderness management* (2nd ed.). Golden: North American Press.

Watson, A. E., Cole, D. N., Tumer, D. L., & Reynolds, P. S. (2000). *Wilderness recreation use estimation: A handbook of methods and systems* (Gen. Tech. Rep. RMRS-GTR-56) (198 pp.). Ogden: U.S. Department of Agriculture, Forest Service, Rocky Mountain Research Station.

Chapter 5
Wild Vistas: Progress in Computational Approaches to 'Viewshed' Analysis

Neil Sang

Abstract Understanding the potential visual impact of landscape change provides the opportunity to both prevent inappropriate change and minimize the impact of necessary service provision. Wilderness areas are, by definition, particularly sensitive to the visible presence of human activity and built structures. As GIS technology has reached disciplines with a specific aesthetic interest in landscape the visual questions posed of data have become more demanding. This may be a matter of the degree of confidence which may be placed on the results of visibility analysis given data uncertainties and vague definitions. Alternatively it may be that more qualitatively subtle questions are to be addressed, concerning not just whether a location is visible, but how prominent it is in the view or indeed how notable it is psychologically. In some cases relevant techniques have existed for decades, but their significance is only now being appreciated. Other aspects are stimulating renewed interest at the research front that bring together disciplines from GIS, computer graphics, landscape planning and psychology. This chapter considers the evolution of the questions being asked of visual analysis as well as techniques and technologies developed to answer them.

Keywords Viewshed • Visual topology • Landscape • Perception • Digital terrain model

5.1 Introduction

Landscape is commonly understood in the Anglophone world to be quintessentially visual (Wylie 2007 p. 55). While acknowledging that other cultures and languages may understand the term to be more proactive, functional or societal in character (Olwig 2002). How a landscape looks is consistently important hence its central role in Environmental Impact Assessment (Wilson 2002). With respect to wilderness landscapes, the term inherently implies a lack of apparent human management. Thus many landscapes may in fact be the product of millennia of farming, but are

N. Sang (✉)
Department of Landscape Architecture, Planning and Management, Swedish University of Agricultural Sciences, Lund, Sweden
e-mail: neil.sang@slu.se

S.J. Carver, S. Fritz (eds.), *Mapping Wilderness*,
DOI 10.1007/978-94-017-7399-7_5

commonly perceived as wild provided they lack built features, or evidently unnatural elements (SNH 2005) such as long straight lines of vegetation which do not respect the underlying terrain. The concept of Wilderness may be lost however once the presence of human activity is indicated, even if that presence is physically relatively small, meaning the task of viewshed analysis for predicting impact of a development in a wilderness is particularly exacting. A telephony mast for example may be barely a filament in the view, but by reminding a visitor that they 'should' check their work e-mail, the spell is broken.

Understanding the potential visual impact of landscape management provides the opportunity to both prevent inappropriate development and allow necessary service provision. As the example of the telephony mast demonstrates however, the subtleties of the issues are demanding more sophisticated answers than simply whether … and possibly 'how much of' [an object] is visible. Indeed, as researchers have addressed these two points, it has become clear that, in fact, neither is at all simple to conceptualise nor to measure. This chapter sets out what progress has been made with respect to understanding the dimensions of the parameter space of a view, and techniques for measuring a view against these dimensions.

Wilderness areas are, by definition, extensive and to some degree inaccessible. Manual monitoring of changes to many vistas within a wilderness is thus at best expensive and at worst impractical or damaging. Virtual models, which allow one to view, and preferably analyse, many vistas remotely are thus desirable. However issues of computation remain significant when dealing with visibility analysis of large areas. Attempts to reduce this problem through lower resolution data, or limiting the number of viewpoints selected raise their own issues, as does the use of areal and remotely-sensed imagery to populate those vistas with land cover information.

There a several dimensions along which this discussion could be structured: computational algorithms; choice of Digital Terrain Model (DTM); application area. Each perspective has slightly different implications for what is considered progress (speed, data integration, analytical richness) Many techniques were first introduced some 20 years ago, yet their significance has not been widely appreciated until recently (see for example the various forms of viewshed implemented by Fisher (1996)) and still task-specific viewshed implementations are hard to find in commercial software.

The structure selected here is to consider the evolution of the techniques along with those of the questions, which have gradually become more demanding. The technology has moved from the research tool of 'Quantitative Geographers', into other professions (archaeology (Wheatley and Gillings 2002; Ogburn 2006; Eve and Crema 2014), planning (Martinez-Falero and Gonzalez-Alonso 1995; Rød and Van Der Meer 2007), ecology (Aspbury and Gibson 2004; Loarie et al. 2013), to eventually merging with computer graphics and gaming technology (De Floriani and Magillo 1996) in order to support virtual worlds (Smelik et al. 2011) and intuitive visualization (Appleton et al. 2002) affording stakeholder involvement (Appleton and Lovett 2005) in interdisciplinary projects which operate with respect to a 'landscape scale' and thus ultimately need to address perceived notions of landscape. Such questions require techniques which model not only that which is visible but that which is perceived, a much wider question (Gibson 1986; Bruce et al. 1996; Fisher 1996; Bishop et al. 2001) which varies with subject, e.g. perceived safety (Jorgensen et al. 2002) or perceived wilderness (Habron 1998; Comber et al. 2010; SNH 2005) and wherein the role of the technique itself must be considered a parameter (Bishop et al. 2001; Bishop and Lange 2005).

5.2 What Is Visible?

5.2.1 Binary in Theory But Fuzzy and Vague in Practice

By far the most common viewshed approach is ray tracing line of site between pixels on a raster Digital Terrain Model (DTM). As generally implemented (Heywood et al. 2010) the algorithm is computationally expensive (i.e. slow) because for every raster cell whose visibility is of interest, a line of sight must be drawn to every other cell in the raster, leading to a steeply increasing processing time with each new cell. For inter-visibility problems the processing time becomes exponential. Processing speed has increased enormously, such that computing an individual viewshed on a small area is relatively quick. It is not, however, yet sufficiently convenient for real time use with stakeholders, use with dynamic view points, nor for multiple viewsheds on massive datasets.

Recently (Carver and Washtell 2012) adapted algorithms from computer gaming engines to allow rapid viewshed analysis of raster DTM. Currently available only as a standalone software, their algorithm makes use of hard wired 3D projection capabilities of Graphics Processing Units to massively decrease the time required per calculation. Rather than calculating a line of sight between each cell and every other cell in the raster, it dynamically divides the DTM along the plane of view, such that cells falling behind and below occluding cells need not be individually calculated (Carver and Washtell 2012). The result is real time viewshed production which not only allows interactive planning with stakeholders, but also raises the possibility of quickly investigating the error margins to the viewshed, and thus any other statistics based upon it.

5.2.2 DTM Error Issues

All DTM provide only an approximation of the height of the terrain at any one location. The underlying Digital Elevation Model (DEM) records height at specific points, be that the point at which a laser bounced from the terrain below (LiDAR), or the point at which a surveyor established their height above a datum level. Today this is likely to be accurate to within a few centimetres thanks to Global Positioning Systems (Heywood et al. 2010), but even this difference can make substantial changes to a viewshed. This is something we intuitively understand; just moving one's head to peer over the edge of a cliff would substantially change our view. For

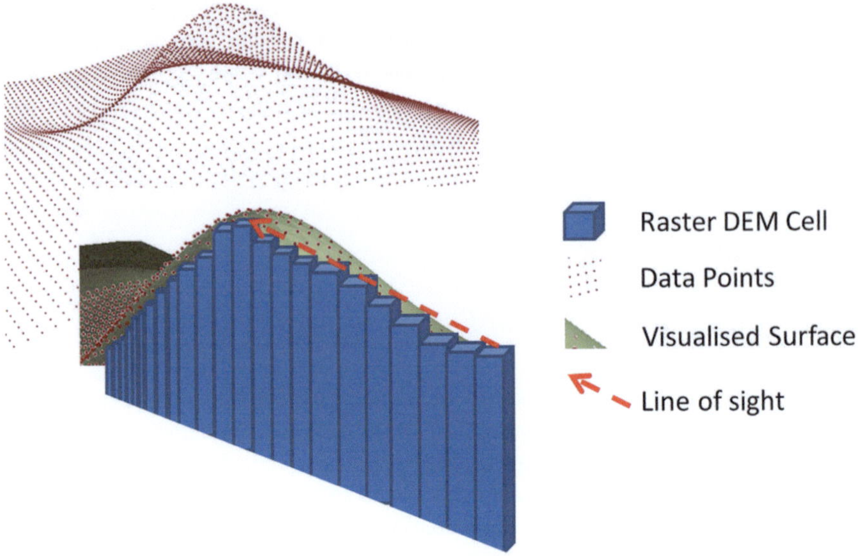

Fig. 5.1 Point heights, digital terrain models and uncertainty in visibility analysis

all but the most exacting problems it is an error margin which is likely to be acceptable but it is important to acknowledge that, just as in real life one would be unlikely to consider an area fully explored without investigating such near view occlusions, so too any single viewshed must be considered only a partial exploration.

Today, the choice of DTM is usually a larger source of potential error than those in the initial measurement. Digital Terrain Models, as the name suggests, are only a model of the terrain, a means of estimating, from the known heights in the DEM, what the likely heights are in the space in between. Raster DTM simply divide the space into squares and give all the locations therein the same height as the data point they are enclosed with. The result is a terrain of 'steps' (Fig. 5.1). The precise location of the viewpoint may place an object just behind the crest of one step but we know that may not be correct because the height of the edge of the step is only a guess based on the height of the point at its centre. Thus for some cells the answer is not really a binary visible or occluded but a 'maybe'. How certain that 'maybe' is depends on the resolution of the raster since a higher resolution raster will have less space between the cell edge and its known centre height. One approach to assessing the significance of such errors is to use a monte carlo simulation, that is to randomly vary the DEM repeatedly and measure the difference in outcome for the viewshed,[1] a simpler approach is to consider the viewshed for several different resolutions of DEM.

[1] An example implementation for ArcGIS can be found here: http://www.arcgis.com/home/item.html?id=5e9cb4fd73fe4288a4cf534cc5a119aa

5.2.3 Digital Surface Models

Errors with respect to the terrain data and model used might be considered 'global' problems, that is to say the problem is the same everywhere, even if relatively more problematic in flat areas or around ridges, where the change in height is very different to the sample rate. A problem which is similar, but considerably more context specific, is that of whether to use a 'ground surface' DTM or a Digital Surface Model (DSM). Theoretically, DTM represent the surface of the earth, but not the vegetation, buildings and so on. The raw data usually registers these surface objects to some degree and thus these are removed to create the DTM. One faces an interesting philosophical issue as to when rock or concrete becomes a building rather than part of the fabric of the terrain. The question is by no means entirely academic. In many parts of the world (e.g. the Lake District National Park, UK), buildings are constructed from local stone and built into the side of the hill, often to minimise visual impact. Whether to 'remove' that building and thus create an entirely spurious hillside or hollow, will have clear implications for both line of sight crossing that location, and the choice of where to place the viewpoint for a viewshed from that dwelling. All of which presupposes a suitable method to identify the 'surface' objects in the first place. Until the advent of LiDAR the point was moot since vegetation could only be crudely approximated and accurate 3D models of the built environment were hard to obtain. LiDAR now presents the opposite problem: how to remove surface features (Priestnall et al. 2000)? Identifying a building as such is not necessarily clear-cut in the absence of accurate cadastral mapping. Distinguishing a slate roof from a slate scree slope, for example, may not be possible by spectral signature, so a considerable manual effort may be required to separate DSM and DTM.

Most DTM contain one particularly important simplification: they assume a topologically simple surface with no holes or bridges. Most 3D modelling in GIS has until recently in fact been only 2.5D, that is a single surface with varying height but no sub-surface data, a hollow terrain. Many algorithms for viewshed mapping assume this to be the case, and cannot be relied upon to accurately map the viewshed of topologically complex 3D models. Most surface triangulations for example assume a 2.5D surface, as do the algorithms for view shed generation which build upon them (Sang 2011).[2]

Provided the objects have solid, opaque, surfaces, complex topological shapes are not necessarily a problem for a ray-traced viewshed, though full 3D voxel analysis precludes many time saving algorithms (e.g. that of Carver and Washtell 2012). For wilderness problems, the issue is primarily how to ensure detailed objects are represented in an otherwise lower resolution terrain, since to model the entire extent at sufficient resolution would often be impractical. A more fundamental problem is

[2]There are some implementations of Voronoin TIN in true 3D but these have yet to have compatible viewshed algorithms developed, e.g. Ledoux H. (2006). Modelling Three-dimensional Fields in Geoscience with the Voronoi Diagram and its Dual, PhD Thesis, University of Glamorgan, November 2006.

posed by visually vague objects, in particular those formed by groups of individual objects that produce a partial occlusion, such as leaves forming tree crowns and trees forming woodland.

For modelling the impact of near view objects and small areas full 3D models have been developed for vegetated environments, but scale of application is necessarily limited. At a larger scale there have been attempts to factor the net transparency of a surface feature into the model (Bartie et al. 2011). These are not methods that are generally applied in current landscape management however, where a more pertinent issue to remember is that height data nearly always originally included surface objects in their raw data and thus potentially significant differences are to be expected between the viewshed based on them and reality. Often DEM based on LiDAR will provide an upper and lower surface height, so the accordance or otherwise of the viewshed from both may provide a degree of confidence that reality has been well approximated.

5.2.4 Observation 'Error' Issues

It is an interesting philosophical question as to whether uncertainty in a viewshed owing to the visual acuity of the human eye may be described as 'error'. The viewshed in this sense is not that to which there is line of sight, but the area within which an object may be seen. Thus line of site may be strictly correct, but the object is so small a part of the view a person cannot see it. Thus the simple ray-traced line of sight viewshed, taking no account of perspective, may be too sensitive for that particular question. The same would not be so if the viewshed were being determined for other purposes such as radio signal reception. Thus if one were interested in siting a telephone mast to maximize reception and minimize visual impact, one would not necessarily use the same viewshed for both cases.

Predicting the viewshed for a particular object to a human observer is considerably more complicated than that for a telephone mast. The acuity of the human eye, in so far as a 'standard' eye may be determined, is only part of the equation. The eye's retina does not register a complete image in the way a digital camera might. The eye is a dynamic actor in perception, actively scanning the scene and able to focus greater attention to some areas as needed (Rensink 2000; Ogburn 2006). Beyond the bio-mechanics of the process there is also a layer of interpretation and psychology (Gibson 1986; Rensink 2000; Brockmole et al. 2012). The balance between innate scene interpretation and learnt scene prioritization is still a matter of considerable debate. Certainly some scenarios will alter the visual salience of an object, breaking the horizon is generally acknowledged to raise salience for example (Fisher 1996; Hernández et al. 2004; Bernardini et al. 2013), making the precise location of the viewpoint a sensitive issue.

It may also be, however, that such subtleties of human vision may be used to better 'camouflage' necessary infrastructure in wilderness areas without recourse to a simple screen of vegetation, which could itself appear out of place. Many principles are already known and understood within landscape planning practice. There have also been attempts to build some of these into visual landscape modelling, however quantification of qualitative understanding is a difficult calculation and the more complex the algorithms for determining and modelling visibility, the longer the process, thus basic computational issues remain at the foundation of automatic scene analysis. For this reason guidelines in landscape planning often determine some visual limit beyond which an object is not considered visually significant (Falconer et al. 2013), however given the apparent importance of context in human perception, one must question the appropriateness of such a simplification to so sensitive an environment as a wilderness landscape.

5.2.5 Fuzzy Viewsheds

Fuzzy viewsheds were first proposed in the early 1990s (Fisher 1994) to address the limitations of the binary viewshed. However implementation in standard GIS software is still unusual. The term 'fuzzy' is based on fuzzy-set theory. Figure 5.2 shows the output from one form of fuzzy viewshed showing the probability of an object being visible based on both line of sight and its apparent height in the view.

The term is also used with respect to viewshed uncertainty stemming from limitations to the acuity of the human eye relative to the size of a viewed object and atmospheric conditions (Ogburn 2006) though these might better be described as

Fig. 5.2 'Fuzzy Viewshed' output from ESRI ArcGIS tool (Rasova 2013) based on a viewed object of 2 m with 100 m maximum clarity (The ArcGIS toolbox tool of this implementation cab be found here: http://www.arcgis.com/home/item.html?id=5e9cb4fd73fe4288a4cf534cc5a119aa)

'vague' viewsheds. Vague viewsheds are highly relevant to visibility modelling in wilderness areas as these are often, at least in Europe and North America, mountainous areas where visibility conditions are highly variable and vegetation seasonal. The viewshed based on 20/20 vision on a clear winter's day where vegetation has little effect may be a rather unrepresentative picture over the course of a year.

5.3 How Much Is Visible?

5.3.1 Vague Viewsheds

Having established that visibility is, in fact, more a question of 'degree' than 'whether', the question then arises as to what a non-binary viewshed might comprise? Whereas a binary viewshed might set a fixed maximum distance of visibility, the visibility may be weighted by distance (relevant say to signal strength for a mobile telephone mast) but the viewed object's apparent size is perhaps more relevant to human vision.

Size decreases due to perspective according to simple trigonometric functions (Equation 5.1):

$$f = \frac{w}{\tan(a)}$$

$$g = \frac{h}{\tan(b)}$$

$$S.x = \frac{P.x * f}{P.z} + \frac{w}{2}$$

$$S.y = \frac{P.y * f}{P.z} + \frac{h}{2}$$

where w = display width, h = display height, a = horizontal angle of view, b = vertical angle of view, $P.x$ = model coordinate, $S.x$ = screen coordinate.

Area thus decreases as the square of the distance function. The Visual Solid Angle (VSA) is the angle extended between one edge of a seen object and another. For example, if a piece of paper is held vertical, facing the viewer then rotated around its centre, the apparent angle between the top and bottom of the paper decreases, as does its apparent area and thus the proportion of the view it takes up.

With respect to some landscapes this is clearly very significant in determining how much of some land covers are visible; Fig. 5.3 shows the effect for visible area of grazed pasture on a plateaux compared with a rocky cliff face when seen from the opposite side of a valley.

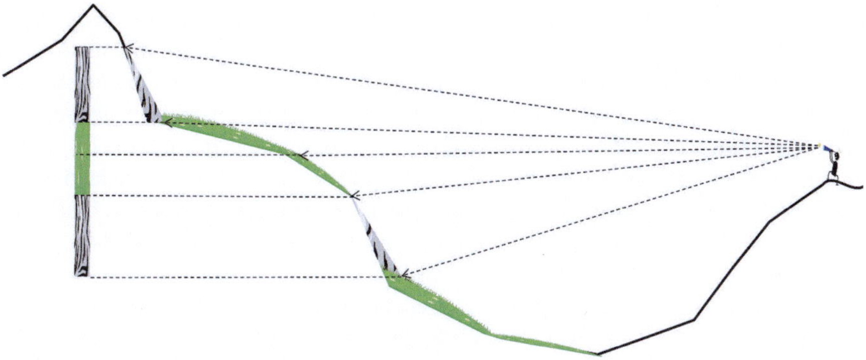

Fig. 5.3 Visual angle and visual prominence of landcover

Chamberlain (2011) used a ray-traced viewshed but weights the visible land covers based on their angle to the viewer, while (Carver and Washtell 2012) uses a viewshed algorithm which directly relies on the VSA.

The choice of DSM or DTM is again critical. Forest edges may form significant blocks in the view but the calculated area based on VSA would be quite different depending on if calculated from a 3D model with vertical sides or terrain model which interpolates vertical change to a medial slope (e.g. a type of DTM base on Triangular Irregular Networks). Full 3D models are computationally expensive for ray tracing a viewshed, particularly if the possibility of holes and tunnels is allowed for, because every point on the surface must be tested. Some proposals have been suggested for measuring degree of visibility through sampling particular points (Rød and Van Der Meer 2007) though this seems more relevant to urban areas (Bartie et al. 2010) where the extent is smaller and the surfaces themselves, usually, solid.

5.3.2 Perspective, Area and Ecological Fallacies

As GIS technology has reached disciplines with a specific aesthetic interest in landscape the visual questions posed of data have become more demanding. In particular visualisation of potential future scenarios is an increasingly popular technique. This involves rendering land cover data into perspective views with photo-realistic representations. In a Virtual Landscape Theatre (Nijnik and Miller 2013) it also entails taking map data at a given scale and rendering it at almost real world size. This is, most likely, a quite different use of the data than that for which it was intended.

Context is important to all classification, both human and machine. For example if one were to look at a wilderness scene showing only rocky mountain tops (Fig. 5.4), an observer might overestimate the presence of rock and underestimate the presence of vegetation in the area. The land cover data, viewed from above, will see a foreshortened cliff face, particularly with respect to the area circled in yellow there would be little information in such a land cover map to indicate the land cover class of the near vertical surface, rather the cliff would likely form a boundary between 'rock' and 'grass' land cover polygons. Depending on the scale of the land cover maps and the degree to which it uses composite classes such as 'rock and grass the data may not be particularly relevant to the perspective view. To assume that the portion of a land cover class visible in a view would have been classified by

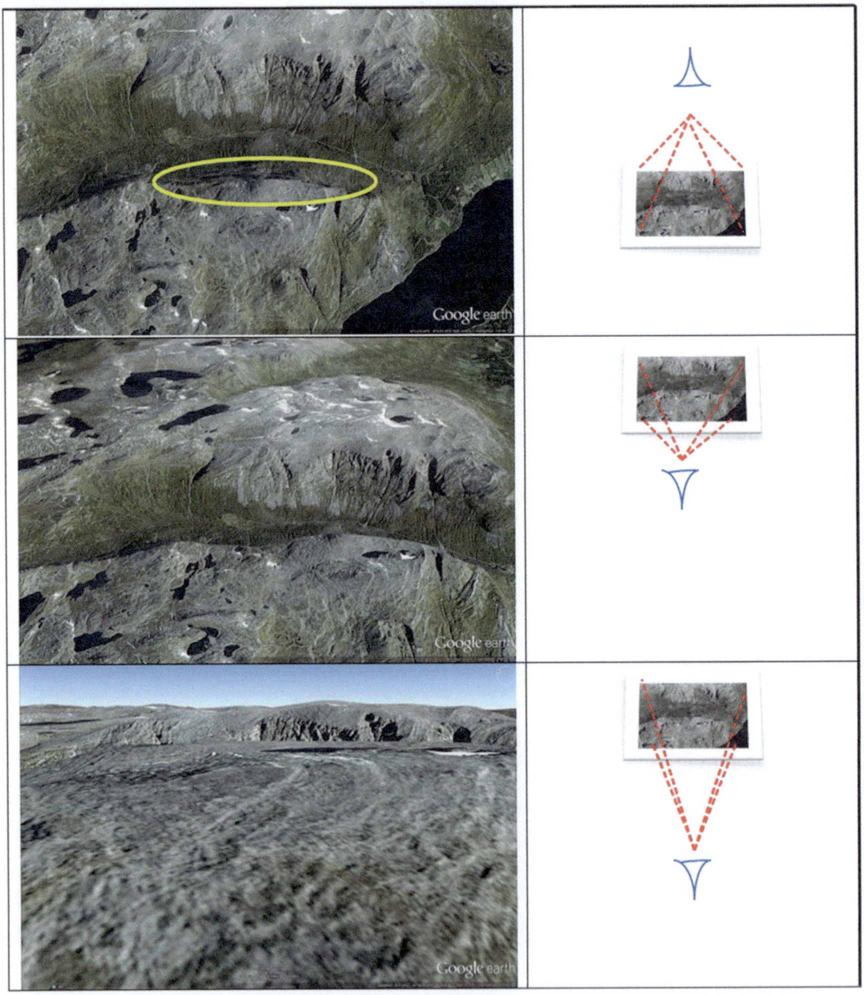

Fig. 5.4 Perspective, scale and apparent land cover class

the surveyor as the same class assigned to the entire polygon, even if it may only be a small fragment of the originally mapped polygon, is to commit an ecological fallacy (Heywood et al. 2010). The same is therefore true for visualizations of that data, a fact which is not, as yet, sufficiently widely recognised.

Visual analysis applies, by definition, to real world scales, and taking a new perspective re-samples the data to a subsample of that originally collected in orthographic view yet renders it to a higher resolution. Thus while it is possible to make calculations as to the apparent area of a terrain visible in different locations, translating that into apparent land cover entails making additional assumptions about the homogeneity of the polygons in the land cover data. An open question, therefore, is how to develop measurements of variance and potential error to reflect the uncertainties within viewshed statistics, uncertainties which are not only due to the fuzzy and vague nature of viewsheds themselves, but also due to the statistical re-sampling of the land cover data which a perspective view introduces?

5.4 What Does It Look Like?

5.4.1 Scene Analysis

For the present, deciding what a scene looks like remains a fundamentally qualitative process. Visibility analysis still has a role in supporting that decision process however. In the first instance it can help reduce the total burden of information to be considered, secondly it can provide some objective measurements to support a qualitative reasoning.

With respect to the first instance, visibility analysis may be required because physically accessing a particular location is impractical. Though by definition this is not likely to be with respect to a human viewer, it is perhaps still relevant to reduce potential visual disturbance to a breeding habitat. More likely the problem is to select those locations which require manual inspection from the numerous potential options. Avoiding the need to physically visit and photograph a particular vista is a significant benefit by itself. However virtually visiting all the potential locations may still be very time consuming. Indeed given the apparently infinite number of possible perspectives onto a particular location, the question arises of how to even compile the first list of possible viewpoints. One motivation for the development of Visual Landscape Indicators (Ode et al. 2008) is thus in order to use a computer to pre-filter out the least relevant cases or find a set of viewpoints which covers the range of landscape characteristics in a representative way.

With respect to supporting the qualitative reasoning of landscape experts charged with characterizing a landscape area, or assessing the likely impact of a landscape change on that area, visibility analysis is a vital pre-requisite because it can limit the area of the map data under consideration to that which corresponds to the real vista in question. How to interpret landscape metrics based on a viewshed with respect to their meaning in a real a vista is not simple however. Patch Richness, for example, being based on visually distinct patches, can be quite different when calculated on the perspective scene to the viewshed because the viewshed may contain 'holes' representing occluded land, while in the scene the two land cover patches lie adjacent and can be visually indistinguishable. Metrics such as Mean Shape, suffer both these topological changes as well as the distortion of shape by perspective (Fig. 5.5).

Figure 5.5 (Sang et al. 2008) demonstrates the degree to which landscape metrics may differ depending on whether they are calculated on the entire map, the viewshed or the scene, even to the point of changing rank, thus calling into question rank correlations with expressed preference. Germino et al. (2001) see this difference as a fundamental problem for relating human perception to cartographic data. Sang et al. (2008) look at whether correlations with expressed preferences are indeed affected by which view of the land cover data is used, they conclude that some metrics are more sensitive to viewpoint change than others, and note that for a few metrics the cartographic map was the better correlate to expressed preference. None-the-less, there is clearly an issue with respect to information consistency and it is desirable to have the ability to also analyse the scene rather than only the viewshed.

Most methods of scene analysis involve capturing the on-screen image. That may be from a photograph or a visualization. For more complex imagery, artificial

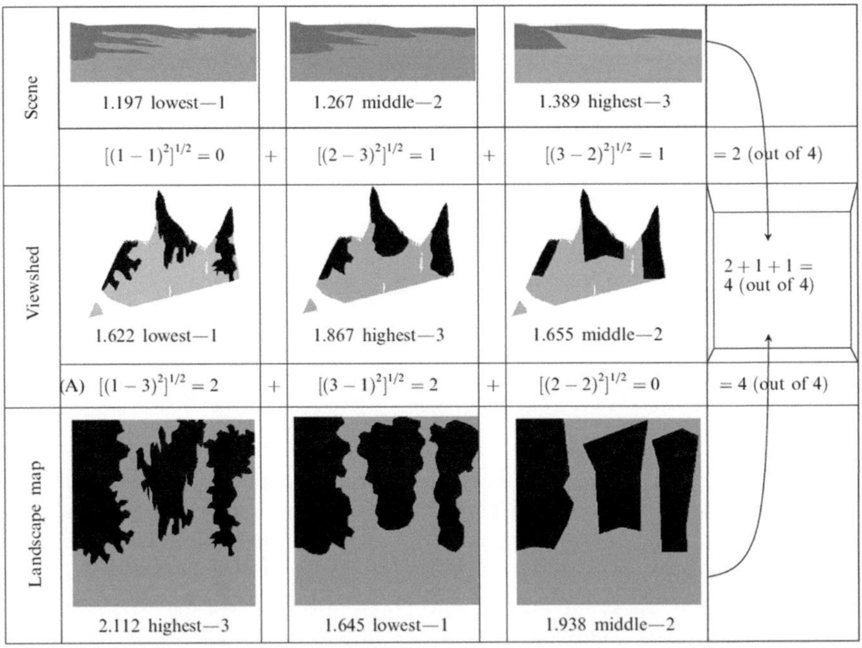

Fig. 5.5 Effect of perspective on landscape metrics (From Sang et al. 2008) Calculation of the rank change matrix in all three view types (with constant, coarse, patch size and varying complexity). Below each image is given SHAPE MN metric score and position. The addition rows display the addition of rank differences: see the discussion in the text for the example of call (A)

intelligence approaches have been developed to segment the scene in to semantic objects. Perhaps best known of these is the software 'e-cognition' (www.ecognition.com), allowing fully automatic or manually trained image segmentation based on colour and pattern. More simple approaches have also been applied by painting different objects in particular colours (Sang, Ode et al. 2008) allowing each patch to be analysed subsequently. All of these approaches however suffer from the same two basic problems:

1. They work on a 'dumb' image. The semantic classifications of the underlying data and their original spatial locations are lost in the graphical projection process, as is the underlying resolution of the data such that they are now only available at the screen resolution.
2. The techniques usually require considerable manual input and are therefore impractical for analyzing large numbers of scenes.

An alternative approach is to embed the viewshed into the original DTM as topological links and thus retain a direct link to the original semantics of the landscape data. Sang (2011) developed a data structure which considers the viewshed as a series of horizons and the shadows they cast on the landscape beyond (after De Floriani and Magillo 1996), and then sets pointers to allow the topology of the scene to be reconstructed. The apparent scene may thus be queried for different land cover scenarios without repeating the viewshed analysis to answer questions such as what will be the effect of change on scene patch richness?

5.4.2 Not All a Matter of Perspective – Visual Invariants

Perhaps the most common and intractable problem for researchers, planners and other managers of the visual landscape, is deciding which viewpoint or set of viewpoints to select. When establishing the visual impact of a new development for example, whether it is seen at great distance or close up will have some effect on its perceived impact. Some views may frame the object, while others serve to partially obscure it. So whether one is assessing a visual impact manually, or attempting to use some landscape metric to quantify it, the perspective(s) selected is clearly important. It is often suggested that one needs to provide a 'representative' range of views, advice which arguably only diffuses the core problem – what is 'representative'? Expert selection of that which is 'representative', risks sample bias either due to that individual's own preferences or simply because they cannot consider all the possibilities. While systematic approaches are able to provide some consistency in characterization of landscape character (SNH 2002), mapping areas according to the character of how it appears from elsewhere (rather than its own internal characteristics) or according to what can be seen from its location, can seem impossible due to the significance of perspective.

A quantitative definition of 'a representative view' runs the risk of bias due to the relevance of the selected metrics to a particular landscape but also due to the fact that there may be few if any scenes which actually contain the 'average' apparent metric as calculated for the area as a whole. More likely is that there may be several distinct characteristic views of the same area, which raises the question of how to quantitatively define and statistically identify them.

Carver et al. (2012), have used their fast viewshed analysis to map each apparent landscape metric and define visual character regions by multivariate clustering. Boundaries to apparent character areas might be expected to emerge where visual metrics quickly change with viewpoint, particularly for example when a new land cover becomes visible. Looking for rapid change in metric value is still based on an understanding of visual character as a continuously varying phenomenon.

An alternative approach is to attempt to break down the view space by that which changes between viewpoints and that which remains the same. Scene invariants are commonly used in computer vision and photogrammetric 3D scene reconstruction to identify the same objects from different viewpoints (Plantinga and Dyer 1990; Van Gool et al. 1995; Ryan et al. 2014). For example scale does not change the ratio of distances between points on an object. The ratio of distances between components on the face would thus be an invariant, which can help identify the same individual from photographs at different distances.

Topology is also used as a visual invariant. The term refers to a branch of mathematics which considers the order or sequential arrangement of things, rather than a scalar metric of distance between them (Kinsey 1991). For example, Fig. 5.6a shows the shadow cast by a torus or 'donut' shape. Between the conical VSA emanating from its centre, the apparent shape of the donut changes, but the hole is not visible. Within either view-cone, the hole is visible. The presence or absence of the hole is thus a local invariant. The difference can be expressed mathematically using graph theory (Kinsey 1991).

Sang (2011) shows that the horizons in a view may be considered as such a graph (Fig. 5.6b). If a viewpoint varies a little the geometry of the horizons will also vary, but we have Equation 5.1 to tell us how that geometry will change. We can predict the likely range of appearance so the idea of an average view can be given some meaningful measure. However, if a change of view completely obscures an existing edge, or reveals a new one, the impact on the scene metrics cannot be predicted.

The topology of the graph will not change unless an occluding edge appears or disappears. So topological change represents one form of invariance which can objectively ascribe some maximum limit to an area beyond which one cannot state that an area has a statistically meaningful 'average' character based on scalar metrics.[3] Figure 5.6c shows how the view space may be divided in zones of different apparent topological complexity (Ode et al. 2010). Within the 1, 2 and 3 horizon zones geometric statistics can meaningfully be said to have some continuous function and thus average and variance to be estimated through random sampling or calculated via a fast viewshed technique such as that of Carver et al. (2012).

[3] Sang et al. (2015) used an online survey to demonstrate that one measure of the topological complexity of the horizon graph was perceptually salient to respondents.

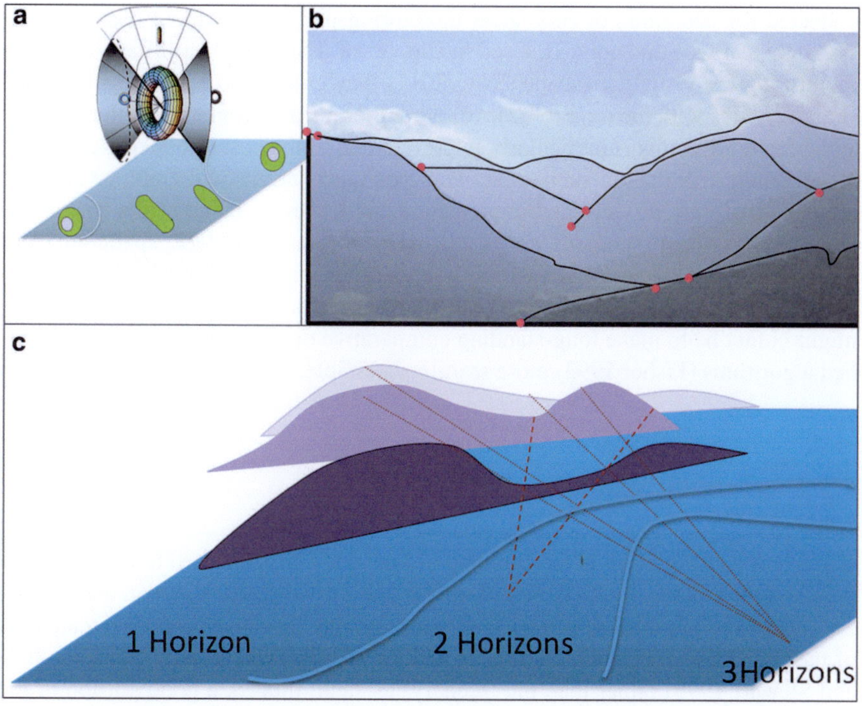

Fig. 5.6 Defining zones where apparent horizon graph topology is invariant

5.5 A Wider Perspective

This chapter has attempted to provide a brief overview of how the whole question of visual analysis has evolved over time as well as some of the techniques developed to answer it. It is by no means a comprehensive history, nor a complete bibliography. Indeed, rather like the subject itself, selection of the questions, techniques and applications is to a considerable degree a matter of perspective depending on whether one is primarily interested in how to best generate of statistics about a particular scene or for an entire area, whether one requires real time visualization or a synopsis with error margins and whether one defines the state of the art as that available only as a prototype or that accessible in practice.

Advancing technical abilities may only serve to highlight broader issues facing planning for visual landscape factors, in particular the availability of the necessary data and the relevance of that data to visualization. Concern about how to evaluate the fuzzy probability that a given land cover may be seen, with a degree of clarity (or vagueness), and assessed as to its apparent extent or salience in the view, from a viewpoint that is partially obscured by topologically complex, partially transparent vegetation is a somewhat 'luxury' concern. Rather these techniques are being applied to mitigate limitations to data which may be at coarse resolutions, or collected for economic or ecological monitoring purposes, without regard for the visual appearance and context. Used in support of a Spatial Data Infrastructure for monitoring the visual landscape (Sang 2011) they could also help integrate visual characteristics with the requirements of other monitoring programs. With respect to Wilderness areas this is particularly important due to the cost of collecting detailed visual land cover data in such areas. These data limitations will continue and it is arguable that current viewshed implementations leave too much of the conceptual definition of the process to end users who, increasingly, lack spatial-analysis training. The visual sensitivity of wilderness areas make these choices potentially significant, thus the most important technical contribution to wilderness planning might in fact be to make long-standing comparative options for the range of viewshed algorithms (Fisher 1993) more readily accessible.

5.5.1 Horizon Scanning or Shoe Gazing?

It is worth bearing in mind the example of one study which aimed to monitor what people were looking at while out hill walking by providing each person with a camera and a fixed schedule of when to take photographs. The results showed that, most of the time, they were looking at near field objects relevant to navigation (Hull Iv and Stewart 1995) rather than the 'scenery' as such. Indeed, a good portion of the time when walking one is actually looking at one's shoes to avoid tripping!

This illustrates an important difference between the way people experience landscape and the way it is generally mapped and monitored. Maps are an effective means of holistically encapsulating the landscape. Monitoring generally aims to be comprehensive or at least representative. Human perception however is highly selective and biased. Thus the very notion of the 'representative view' is in some regards an illusion. At the very least we need to be directing the attention of our visual analyses, and our land use planning and management processes, not simply onto that which can be seen, but rather to that which is likely to be noticed. In this respect the use of eye tracking technology to help infer what about a scene draws the attention (Pihel et al. 2014) will be equally relevant to future scene analysis as techniques to simulate such perception.

What is likely to be noticed is not simply defined by relative scale, colour contrast or geometric framing but also by context. Wilderness is particularly sensitive to the visual impact of human development because, by definition, it is out of context. The human eye is tremendously acute – in theory it can see a candle flame from 48 km – so even very small changes will be seen. Moving from analysing 'whether and possibly how much of' something can be seen, to what it looks like, is thus vital to monitoring and mitigating visual impact in wilderness areas.

Acknowledgements The author would like to acknowledge the contributions of Chris Gold (Prof. Emeritus, University of Glamorgan) and David Miller (James Hutton Institute), who supervised the PhD work referred to here as Sang 2011. Discussions with Åsa Ode Sang and Caroline

Hägerhäll have also been helpful in developing the ideas expressed here. This chapter is dedicated to the memory of Professor Peter Fisher, a kind and inspiring teacher who made a great contribution to Geo-Information Science and to visibility analysis in particular.

References

Appleton, K., & Lovett, A. (2005). GIS-based visualisation of development proposals: Reactions from planning and related professionals. *Computers, Environment and Urban Systems, 29*(3), 321–339.

Appleton, K., Lovett, A., et al. (2002). Rural landscape visualisation from GIS databases: A comparison of approaches, options and problems. *Computers Environment and Urban Systems, 26*, 141–162.

Aspbury, A. S., & Gibson, R. M. (2004). Long-range visibility of greater sage grouse leks: A GIS-based analysis. *Animal Behaviour, 67*(6), 1127–1132.

Bartie, P., Reitsma, F., et al. (2010). Advancing visibility modelling algorithms for urban environments. *Computers, Environment and Urban Systems, 34*(6), 518–531.

Bartie, P., Reitsma, F., Kingham, S., & Mills, S. (2011). Incorporating vegetation into visual exposure modelling in urban environments. *International Journal of Geographical Information Science, 25*, 851–868.

Bernardini, W., Barnash, A., et al. (2013). Quantifying visual prominence in social landscapes. *Journal of Archaeological Science, 40*(11), 3946–3954.

Bishop, I., & Lange, L. (2005). Vizualisation Classified. In Bishop, I., & Lange, E. (Eds.). *Vizualisation in landscape and environmental planning: Technology and applications.* Taylor and Francis: New York.

Bishop, I. D., Ye, W.-S., & Karadaglis, C. (2001). Experiential approaches to perception response in virtual worlds. *Landscape and Urban Planning, 54*(1), 117–125.

Brockmole, J. R., Davoli, C. C., & Cronin, D. A. (2012). The visual world in sight and mind: How attention and memory interact to determine visual experience. In B. H. Ross (Ed.), *The psychology of learning and motivation* (Vol. 57, pp. 103–145). London: Elsevier.

Bruce, V., Green, R., et al. (1996). *Visual perception, physiology, psychology and ecology*. Hove: Psychology Press.

Carver, S., & Washtell, J. (2012). *Real-time visibility analysis and rapid viewshed calculation using a voxel-based modelling approach.* GIS research UK 2012, April 11th–13th, Lancaster, UK.

Carver, S., et al. (2012). A GIS model for mapping spatial patterns and distribution of wild land in Scotland. *Landscape and Urban Planning, 104*(3–4), 395–409.

Chamberlain, B. (2011). *Toward human-centered approaches in landcape planning; exploring geospatial and visualisation techniques for the management of forest aesthetics*. PhD thesis, University of British Colombia, Vancouver.

Comber, A., Carver, S., et al. (2010). Different methods, different wilds: Evaluating alternative mappings of wildness using fuzzy MCE and dempster-shafer MCE. *Computers, Environment and Urban Systems, 34*(2), 142–152.

De Floriani, L., & Magillo, P. (1996). Representing the visibility structure of a terrain through a nested horizon map. *International Journal of Geographical Information Systems, 10*(5), 541–562.

Eve, S. J., & Crema, E. R. (2014). A house with a view? Multi-model inference, visibility fields, and point process analysis of a bronze age settlement on Leskernick Hill (Cornwall, UK). *Journal of Archaeological Science, 43*, 267–277.

Falconer, L., Hunter, D.-C., et al. (2013). Visual, seascape and landscape analysis to support coastal aquaculture site selection. *Land Use Policy, 34*, 1–10.

Fisher, P. (1993). Algorithm and implementation uncertainty in viewshed analysis. *International Journal of Geographical Information Systems, 7*(4), 331–347.
Fisher, P. (1994). Probable and fuzzy models of the viewshed operation. In M. F. Warboys (Ed.), *Innovations in GIS 1*. London: Taylor & Francis.
Fisher, P. (1996). Extending the applicability of viewsheds in landscape planning. *Photogrammetric Engineering & Remote Sensing Journal Of The American Society For Photogrammetry And Remote Sensing, 62*(11), 1297.
Germino, M., Reiners, W., et al. (2001). Estimating visual properties of Rocky mountain landscapes using GIS. *Landscape and Urban Planning, 53*, 71–83.
Gibson, J. (1986). *The ecological approach to visual perception*. Hillsdale: Erlbaum Associates.
Habron, D. (1998). Visual perception of wild land in Scotland. *Landscape and Urban Planning, 42*(1), 45–56.
Hernández, J., García, L., et al. (2004). Assessment of the visual impact made on the landscape by new buildings: A methodology for site selection. *Landscape and Urban Planning, 68*(1), 15–28.
Heywood, I., Cornelius, S., & Carver, S. (2010). *An introduction to geographical information systems*. Harlow: Prentice-Hall.
Hull Iv, R. B., & Stewart, W. P. (1995). The landscape encountered and experienced while hiking. *Environment & Behavior, 27*(3), 404–426.
Jorgensen, A., Hitchmough, J., et al. (2002). Woodland spaces and edges: Their impact on perception of safety and preference. *Landscape and Urban Planning, 60*, 135.
Kinsey, C. (1991). *Topology of surfaces*. New York: Springer.
Ledoux, H. (2006, November). *Modelling three-dimensional fields in geoscience with the voronoi diagram and its dual*. PhD thesis, University of Glamorgan.
Loarie, S. R., Tambling, C. J., et al. (2013). Lion hunting behaviour and vegetation structure in an African savanna. *Animal Behaviour, 85*(5), 899–906.
Martinez-Falero, E., & Gonzalez-Alonso, S. E. (1995). *Quantitative techniques in landscape planning*. New York: Lewis Publishers.
Nijnik, M., & Miller, D. (2013). Chapter 25 – Targeting sustainable provision of forest ecosystem services with special focus on carbon sequestration. *Developments in Environmental Science.* N. C. P. C. T. N. M. J. P. T. G. W. R. Matyssek and E. Paoletti, Elsevier. *13*, 547–568.
Ode, Å., Tveit, M., et al. (2008). Capturing landscape visual character using indicators: Touching base with landscape aesthetic theory. *Landscape Research, 33*(1), 89–117.
Ode, Å., Hägerhäll, C., et al. (2010). Analysing visual complexity: Theory and application. *Landscape Research, 35*(1), 111–131.
Ogburn, D. (2006). Assessing the level of visibility of cultural objects in past landscapes. *Journal of Archaeological Science, 33*, 405–413.
Olwig, K. (2002). *Landcape, nature and the body politic: From Britiain's renaissance to America's new world.* Madison/Wisconsin: University of Wisonsin Press.
Pihel, J., Ode-Sang, Å., Hägerhäll, C., Nyström, M., & Sandström, E. (2014, May). Assessments and eye movements compared between photographs and visualizations of logged forest vistas. For what kind of assessments are visualisations a good representation of photographs? Girot/Grêt-Regamey/Wissen Hayek/Fricker/Buhmann (Eds.). *Peer reviewed proceedings digital landscape architecture 2014*, ETH Zurich, Swiss Federal Institute of Technology, Wichmann Verlag Offenbach Berlin.
Plantinga, H., & Dyer, C. (1990). Visibility occlusion and the aspect graph. *International Journal of Computer Vision, 5*(2), 136–160.
Priestnall, G., Jaafar, J., et al. (2000). Extracting urban features from LiDAR digital surface models. *Computers, Environment and Urban Systems, 24*(2), 65–78.
Rasova, A. (2013). *ANALÝZA VIDITEĽNOSTÍ V PROSTREDÍ GIS – Visibiluty analysis in a GIS enviornment*. http://www.kgk.sk/fileadmin/templates/downloads/Zborn%C3%ADk_refer%C3%A1tov_21._SGD/5_4_Rasova_p.pdf
Rensink, R. A. (2000). Seeing, sensing, and scrutinizing. *Vision Research, 40*(10–12), 1469–1487.

Rød, J. K., & Van Der Meer, D. (2007). Visibility and dominance analysis: Assessing a high building project in Trondheim. In *Proceedings, ScanGIS*, Aas, Norway.

Ryan, D., Denman, S., Fookes, C. B., & Sridharan, S. (2014). Scene invariant multi camera crowd counting. *Pattern Recognition Letters, 44*, 98–112.

Sang, N. (2011). *Visual topology: A spatial data structure for modelling visual landcape perception*. PhD thesis, University of Glamorgan, UK.

Sang, N., Ode, Å., et al. (2008). Landscape metrics and visual topology in the analysis of landscape preference. *Environment and Planning B: Planning and Design, 35*(3), 504–520.

Sang, N., Hagerhall, C., & Ode, S. (2015). The Euler character: Aÿnew type of visual landscape metric? *Environment and Planning B: Planning and Design, 42*, 110–132.

Smelik, R. M., Tutenel, T., et al. (2011). A declarative approach to procedural modeling of virtual worlds. *Computers & Graphics, 35*(2), 352–363.

SNH. (2002). *Landscape character assessment guidance for England and Scotland.* http://www.snh.org.uk/wwo/sharinggoodpractice/CCI/cci/guidance/Main/Content.htm#full

SNH. (2005). Public Perception of Wildness in Scotland Report for Loch Lomond & The Trossachs National Park Authority, Cairngorms National Park Authority & Scottish Natural Heritage In Association With Research Now July 2012.

Van Gool, L., Moons, T., et al. (1995). Vision and Lie's approach to invariance. *Image and Vision Computing, 13*(4), 259–277.

Wheatley, D., & Gillings, M. (2002). *Spatial technology and archaeology: The archaeological applications of GIS*. London: Taylor and Francis.

Wilson, S. (2002). Guidelines for landscape and visual impact assessment. In *The Landscape Institute and Institute of Environmental Management and Assessment*. London: Spon Press.

Wylie, J. (2007). *Landscape*. London: Routledge.

Chapter 6
Mapping Human Impact Using Crowdsourcing

Linda See, Steffen Fritz, Christoph Perger, Christian Schill, Franziska Albrecht, Ian McCallum, Dmitry Schepaschenko, Marijn Van der Velde, Florian Kraxner, Ujjal Deka Baruah, Anup Saikia, Kuleswar Singh, Sergio de-Miguel, Rubul Hazarika, Ankita Sarkar, Abel Alan Marcarini, Mrinal Baruah, Dhrubajyoti Sahariah, Trishna Changkakati, and Michael Obersteiner

Abstract This chapter outlines how crowdsourcing and Google Earth have been used to create the first global crowdsourced map of human impact. Human impact in this context refers to the degree to which the landscape has been modified by humans as visible from satellite imagery on Google Earth. As human impact is measured on a continuum, it could be used to indicate the wildest areas on the Earth. This bottom-up approach to mapping using the crowd is in contrast to more traditional GIS-based wilderness mapping methods, which integrate proximity-based layers of remoteness and indicators of biophysical naturalness in a top-down manner. Data on human impact were collected via a number of different data collection campaigns using Geo-Wiki, a tool for visualization, crowdsourcing and validation of global land cover. An overview of the crowdsourced data is provided, along with the resulting map of human impact and a visual comparison with the map of human

L. See (✉) • S. Fritz • C. Perger • C. Schill • F. Albrecht • I. McCallum • D. Schepaschenko
M. Van der Velde • F. Kraxner • M. Obersteiner
International Institute of Applied Systems Analysis (IIASA), Laxenburg, Austria
e-mail: see@iiasa.ac.at

U.D. Baruah • A. Saikia • K. Singh • A. Sarkar • M. Baruah • D. Sahariah
T. Changkakati
Gauhati University, Guwahati, Assam, India

R. Hazarika
B. Borooah College, Guwahati, Assam, India

S. de-Miguel
Faculty of Science and Forestry, University of Eastern Finland, P.O. Box 111, 80101, Joensuu, Finland

Departament de Producció Vegetal i Ciència Forestal, Universitat de Lleida-Agrotecnio Center (UdL-Agrotechnio), Av. Rovira Roure, 191, E-25198 Lleida, Spain

A.A. Marcarini
AgroParis Tech, Nancy, France

S.J. Carver, S. Fritz (eds.), *Mapping Wilderness*,
DOI 10.1007/978-94-017-7399-7_6

footprint (Sanderson EW, Jaiteh M, Levy MA, et al. BioSci 52:891–904. doi:10.1641/0006-3568(2002)052[0891:THFATL]2.0.CO;2, 2002).

Keywords Wilderness • Human impact • Human footprint • Geo-Wiki • Crowdsourcing • Land cover

6.1 Introduction

The large and ever-increasing influence of humans on the Earth's ecosystems is acknowledged as an important driver of environmental change, particularly through continued deforestation, large-scale land acquisition and the expansion of agriculture. Areas of true wilderness are diminishing as humans encroach upon more of the Earth's surface, which has motivated the need to map the spatial distribution of wilderness or wild land areas globally, regionally and at the national level. As there is no agreed-upon definition for what constitutes wilderness (Applet et al. 2000), many of the attempts to map wilderness have combined multiple input layers that reflect two main concepts: (i) remoteness from human influence and (ii) naturalness, which were set out originally by Lesslie et al. (1993) as part of a comprehensive inventory of wilderness undertaken by the Australian Heritage Commission. In particular, the approach considered remoteness from settlements; remoteness from access, i.e. road and rail networks; naturalness of the landscape in terms of the degree to which it is free from buildings and other permanent structures such as electricity pylons; and biophysical naturalness, which is the degree to which an area is free from modern technological society, e.g. through disturbance of the vegetation (See Chap. 2). Other wilderness mapping inventories have used similar approaches, e.g. mapping the wilderness of the Arctic Barents region (Henry and Husby 1995). Modifications have also been made for area-specific studies, e.g. the addition of factors that characterize ruggedness or the physically challenging nature of the terrain in Scotland (Carver et al. 2012). In mapping the wilderness of the United States, population density was added as a way of characterizing solitude or remoteness from permanent inhabitants, the number of dams was considered to be a reflection of uncontrolled processes, and pollution and night time lights were used to develop indicators of naturalness (Applet et al. 2000). At the global level, an evaluation of wilderness was undertaken by the Sierra Club and the World Bank (McCloskey and Spalding 1989) while more recent efforts have attempted to reflect anthropogenic influences even more heavily, e.g. in the mapping of the human footprint by Sanderson et al. (2002) and by Ellis and Ramankutty (2008) in their classification of anthropogenic biomes, which is based on input layers of population density, land cover and land use.

This chapter presents an entirely different approach to mapping wilderness using the concept of human impact. With the help of Geo-Wiki, which is a visualization, crowdsourcing and validation tool for improving global land cover (Fritz et al. 2012), volunteers were asked to identify the degree of human impact (on a scale from 0 to 100 %) which is visible from Google Earth imagery; the concept is explained in more detail in section 5.2. Through different Geo-Wiki crowdsourcing competitions, more than 150,000 samples of human impact were collected globally at more than 100,000 unique locations. An overview of this dataset is provided, which was then used to create a map of human impact using simple interpolation; this approach draws upon a simplified remoteness concept, i.e. distance to visible human influence. The implications and limitations of such an approach are discussed along with plans for future research.

6.2 The Concept of Human Impact

Theobald (2004) developed a human modification framework that characterizes landscapes based on two criteria, i.e. how natural versus artificial the landscape is and how free versus controlled the natural processes are that act upon a particular landscape. Wilderness is at the two extremes of these criteria, i.e. natural landscapes with little evidence of human influence such as settlements or roads, which are characterized by natural processes that are devoid of control, e.g. wild fires that are allowed to occur freely or a low density of dams within a watershed. At the opposite end of the two spectra are urban areas that are highly artificial where natural processes are heavily controlled, e.g. presence of flood defences, urban gardening and landscaping. In between these two extremes are different kinds of landscapes that have been modified by humans to differing degrees, e.g. croplands, rangelands, ex-urban areas, etc.

The framework of Theobald (2004) has been used to guide the concept of human impact as used in this research. Human impact in our context refers specifically to evidence of human modification of the landscape that can be seen from Google Earth imagery, captured as a value between 0 and 100 %. Table 6.1 provides an overview of the gradient of human impact where 0 % indicates no evidence of human activity and would be the wildest landscapes visible, 100 % would be urban

Table 6.1 Overview of human impact

Human impact	Description
0 %	No evidence of any human activity visible
1–50 %	Some visible evidence of human activities such as tracks/roads; evidence of managed forests; some evidence of deforestation; some scattered human dwellings, some scattered agricultural fields; some evidence of grazing
51–80 %	Increasing density of agriculture from subsistence on the lower end to intensive, commercial agriculture with large field sizes on the upper end
81–99 %	Urban areas with decreasing amounts of green space and increasing density of housing
100 %	A built up urban area with no green space, typically the business district of a city

Source: See et al. (2013)

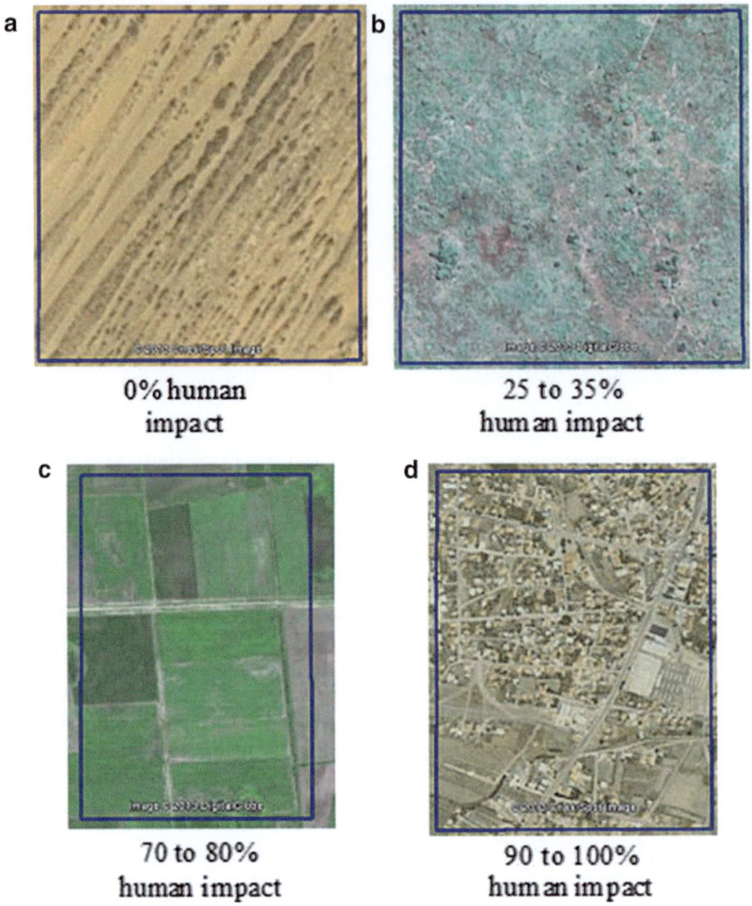

Fig. 6.1 Examples of different degrees of human impact as shown on Google Earth: (**a**) No evidence of human impact; (**b**) some evidence of human impact in an area of largely tree cover; (**c**) an intensively cultivated area with human impact between 70 and 80 %; and (**d**) an urban landscape between 90 and 100 % human impact

areas with no visible green space, and other types of modification would be located along this gradient. Managed forests, or forests with evidence of tracks and some deforestation, indicate increasing human impact followed by grazing and rangelands. Croplands are located in the upper half of the human impact scale depending on field size and intensity followed by urban areas, which have varying types of land use. Thus, suburban areas with green spaces have a lower human impact than fully built-up areas. These types of landscapes are characterized similarly by Theobald (2004) within his framework.

Figure 6.1 provides examples of different landscapes that are visible from Google Earth and their corresponding degree of human impact from an area of wilderness in a desert landscape to a highly built up urban area.

6.3 The Human Impact Dataset

This section describes how the data were collected using Geo-Wiki and then examines how human impact varies across different land cover types and by competition. A comparison of the crowdsourced data with a set of control values, i.e. a set of samples where experts agreed upon the value of the human impact in order to assess the quality of the crowd, is also presented.

6.3.1 Collection of the Data Via Geo-Wiki

The data on human impact (as well as land cover) were collected during four different Geo-Wiki campaigns where each one had a different theme or research question that drove the data collection competition and hence the geographical sampling of pixels on Google Earth. The first competition was driven by the need to validate a map of land availability for biofuel production (Perger et al. 2012; Fritz et al. 2013) while the second one was focused on sampling values from areas where global land cover maps currently disagree (Fritz et al. 2011). The third competition was aimed at specifically collecting human impact and land cover at points on the Earth that correspond to the same locations as those used to validate a new 30 m global land cover map[1] (FROM-GLC) produced by Tsinghua University (Gong et al. 2013) while the final competition was run at these same locations with the purpose of building up a robust crowdsourced dataset for the validation of land cover products more generally. Figure 6.2 shows the Geo-Wiki data collection interface from the third competition where users were asked to indicate the three main land cover types and the human impact within a 1 km pixel shown by the dark outline.

6.3.2 An Overview of the Crowdsourced Data

Over the four competitions, 151,942 validation points were collected (Table 6.2). In some competitions, the participants were provided with the same set of control points in order to monitor their performance. Thus there are a small number of locations for which many validations are available. In other competitions we wanted the same location to be validated at least twice. For these reasons the total number of unique locations for which there is information on human impact and land cover type is 103,509.

There is a highly skewed distribution in terms of the number of validation points contributed by the different users as shown in Fig. 6.3, i.e. of the 1500 registered Geo-Wiki users, only a small percentage contributed the vast majority of validations.

[1] The land cover map can be downloaded from: http://data.ess.tsinghua.edu.cn/

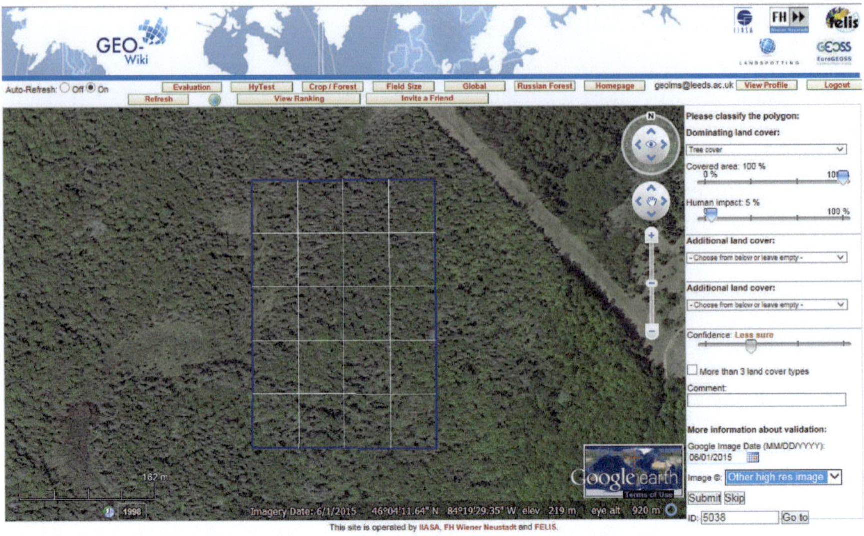

Fig. 6.2 A screenshot from the third Geo-Wiki campaign to collect data on human impact in order to characterize wilderness extent. This particular competition included a tutorial at the start to help train participants

Table 6.2 Number of data points collected by each competition

Competition	Number of contributions	Number of unique locations
1	53,278	33,815
2	30,359	8571
3	32,861	27,814
4	35,444	33,309
Total	151,942	103,509

This is partly due to the nature of the prizes awarded, i.e. in most competitions the prize was co-authorship for the top 10 validators based on a combination of quantity and quality. Thus the number of volunteers per competition was on the order of 50 while in competition four, nine students were paid to collect the data.

In competition 1, the volunteers were asked to identify the dominant land cover type from one of 10 simple classes: (1) Tree cover; (2) Shrub cover; (3) Herbaceous vegetation/Grassland; (4) Cultivated and managed; (5) Mosaic of cultivated and managed/natural vegetation; (6) Flooded/wetland; (7) Urban; (8) Snow and ice; (9) Barren; and (10) Open Water, and to then assess the human impact across the entire 1 km^2 pixel. Feedback from competition 1 indicated that volunteers found it hard to determine a dominant land cover type in numerous cases and that specifying the percentage of different land cover types across the pixel would be easier. As a result the Geo-Wiki interface was changed and in competitions 2 to 4, volunteers entered up to 3 land cover types and the percentage of each one. Then for each of these different land cover types, a value for human impact was entered separately.

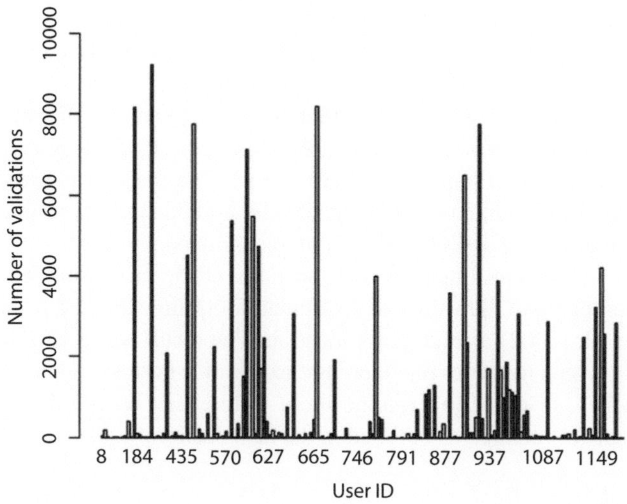

Fig. 6.3 Number of contributions by participant across all competitions

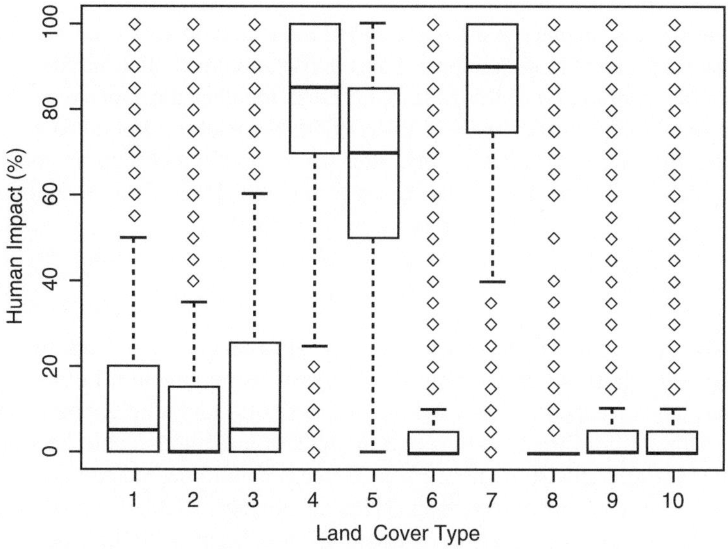

Fig. 6.4 Distribution of human impact across different land cover types: (*1*) Tree cover; (*2*) Shrub cover; (*3*) Herbaceous vegetation/Grassland; (*4*) Cultivated and managed; (*5*) Mosaic of cultivated and managed/natural vegetation; (*6*) Flooded/wetland; (*7*) Urban; (*8*) Snow and ice; (*9*) Barren; and (*10*) Open Water

The overall human impact for each validation point was then calculated as a weighted average of the individual human impact values based on the percentage of each land cover type occupying the pixel.

Figure 6.4 shows the human impact across each of the ten land cover types where the results are as expected, i.e. land cover types (4) Cultivated and managed, (5)

Mosaic of cultivated and managed/natural vegetation and (7) Urban are all on the higher end of human impact while the rest, which are generally more natural land cover types, have much lower values of human impact.

Manual checking of some of the outliers revealed examples of what were often very complex landscapes, particularly in competition 1 where participants were required to choose the dominant land cover type. The result was higher values of human impact than might have been expected for certain land cover types but which reflected human impact visible in these complex, mixed pixels.

The values of human impact at the locations of the crowdsourced data points were then compared to the values of the human footprint extracted from the map of Sanderson et al. (2002) at each point location. The result was a correlation coefficient of 0.56, which indicates that there is clearly some correspondence between the two approaches. In order to examine these differences, a crowdsourced map of human impact was created as outlined in the next section.

6.4 A Crowdsourced Map of Human Impact

A simple inverse distance weighted interpolation method was used to create the crowdsourced map of human impact. This interpolation method is based on Tobler's first law of geography, i.e. things that are close together are more related to one another than things further away (Tobler 1970). For each grid point to be interpolated, the algorithm identifies all the other points within a certain neighbourhood and calculates a weighted vector, w, based on a simple inverse power function:

$$w(d) = \frac{1}{d^x} \tag{6.1}$$

where d is the distance and x governs the rate of distance decay. Each interpolated point is then calculated as a weighted average of its neighbours. In this study the default values in ArcGIS were used, i.e. a power of 2 and a neighbourhood of 12 points. Although different settings and interpolation methods could be employed, the point was to demonstrate how a simple interpolation method can effectively be used to create a crowdsourced map of human impact, which is shown below in Fig. 6.5. No attempt was made at this point to experiment with different interpolation algorithms or the default settings in ArcGIS but this will be undertaken in further research using this dataset.

From this Fig. 6.5 one can see areas of high human impact in the agricultural belts of Canada, the USA and Brazil as well as the big cities on the western coast of the USA and in Mexico. Likewise there is high human impact across most of Europe, India, eastern China, and along the coastal fringes of Australia and North Africa, which reflects areas of high population density. However, looking across Africa as a continent, there is much less human impact evident. Where there are pockets of higher human impact, they clearly reflect locations of agricultural areas

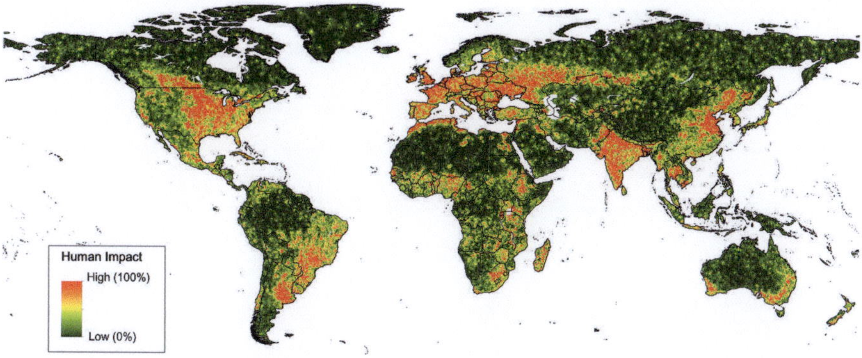

Fig. 6.5 The degree of human impact interpolated from pixels that were interpreted by the crowd using Google Earth

and urban centres. Interestingly, Madagascar shows considerable evidence of human impact throughout the island, which is in line with the relatively small amounts of rainforest left. Large areas of lower human impact coincide with deserts, large areas of tropical rainforest and the temperate forests of the northern latitudes as well as the tundra. Although this would be expected, these areas are also where the lowest resolution satellite imagery is available on Google Earth, i.e. the base Terrametrics imagery at a resolution of 15 m. Thus, evidence of human impact can be difficult to see clearly on some of these images. These areas of low human impact also show the artificial effects of the interpolation where the sampled areas are shown as small pockets across an area that looks otherwise to be devoid of human impact. With a much greater sample size, the visible effect of these artefacts would be minimized and a smoother transition would be produced. Despite these limitations, the crowd-sourced map appears to reflect an overall picture of human impact that conforms with areas of human habitation and activity.

These spatial patterns can be compared with the map of human footprint (Sanderson et al. 2002), which is also on a scale of 0 to 100 % and is shown below in Fig. 6.6.

The Sanderson map was developed using a top-down approach whereby global datasets were combined into a single indicator of human impact. Four categories of input data were used: gridded population density, land transformation (based on existing maps of land cover, built up areas and settlements), accessibility (based on access to roads, rivers and coastlines) and presence of electricity infrastructure (determined through remotely sensed data on night time lights). These layers were combined to produce a human influence index, which is based on the concept of 'remoteness' from existing anthropogenic influences. As outlined previously, this is a commonly used approach to determine areas of wilderness. Sanderson et al. (2002) then normalized the human influence index to account for the presence of different biomes and produced a spatial distribution of the human footprint on a scale from 0 to 100.

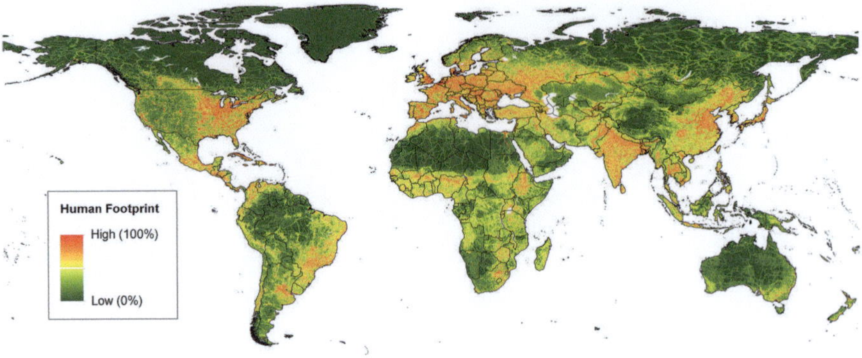

Fig. 6.6 Map of human footprint produced by Sanderson et al. (2002)

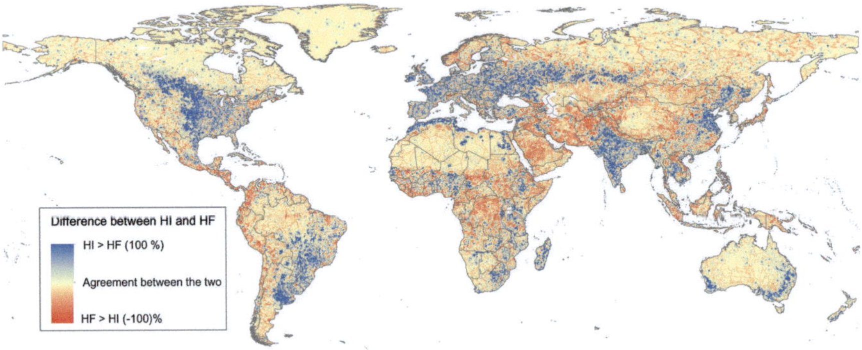

Fig. 6.7 Difference between crowdsourced map of human impact and the map of human footprint from Sanderson et al. (2002)

The maps in Figs. 6.5 and 6.6 have the same scales so it is immediately clear that the amount of human impact in the Sanderson et al. map (2002) is much lower than the crowdsourced map and that the areas with highest human impact are less widespread. However, there is a clearly an overall agreement between the two maps with many of the agricultural and urban areas corresponding to the areas with highest human impact. In contrast, because of the approach used by Sanderson et al. (2002), the map of human footprint shows road patterns and national/sub-national borders as a result of some of the input datasets used, unlike the much smoother pattern shown by the crowdsourced map. These border effects are a result of the input datasets that they used.

In order to compare these two different approaches in a more quantitative way, the maps were subtracted from one another to create a difference image as shown in Fig. 6.7. The blue shading shows areas where the crowdsourced map shows higher human impact compared to the human footprint, yellow denotes areas of agreement while red areas show where the map of human footprint indicates areas of greater human impact than the crowdsourced map.

Figure 6.7 shows that there are large differences primarily in areas of agriculture where there are higher values of human impact using the crowdsourced approach compared to the human footprint. Since the human footprint is based partly on gridded population density, there may be many areas where density is low in rural areas but Google Earth shows a different picture. The presence of road networks and borders also show up as visible differences between the two images. Moreover, areas of Africa, China and the Middle East all have a higher human impact according to Sanderson et al. (2002) compared to the crowdsourced map. Google images suggest landscapes that are less influenced by humans than that which results from a more top-down methodology. Some of these differences may arise because of the fact that the two approaches may actually be measuring slightly different concepts, e.g. assessing human impact via Google Earth considers the direct impact of man-made features on the Earth's surface but it does not directly take remoteness from anthropogenic features into account, as with more traditional wilderness mapping.

Each approach to mapping human impact clearly has advantages and disadvantages. In the case of the crowdsourced approach, the data can be collected very easily but this approach relies on only a sample of data, where some areas may need a much denser representation to accurately reflect human impact. Scaling up this approach to produce a truly representative map of human impact may require far more data than have currently been collected. Moreover, the data have been collected at a resolution of 1 square km. As this resolution contains many heterogeneous pixels of mixed land cover types, determination of human impact is complicated. Increasing the resolution of the sample may improve the ease with which human impact can be identified although this will result in a trade-off in terms of how much data can be collected.

There are also issues which arise regarding the quality of crowdsourced data. Although control points of known human impact have been used to determine overall quality, this applies only to a very small number of points. More systematic methods of bias correction and more interactive and ongoing crowd training need to be incorporated into future crowdsourcing campaigns. The temporal element of Google Earth images is another issue, where images are available from different time periods. However, we have started to collect this information so that we can use it to filter out data from old images or provide some indication of certainty based on the currency of the data.

In contrast the map of human footprint was created using global datasets that have comprehensive spatial coverage. Yet Sanderson et al. (2002) clearly acknowledge that there are potential problems with these datasets. Land cover data have been shown to have high spatial uncertainties and accuracies that even today are still only between 65 and 75 % (Fritz et al. 2011; Gong et al. 2013). There are also issues with all of the other datasets used, as pointed out by Sanderson et al. (2002), e.g. potential incompleteness of the road data, and gridded population data have problems with accuracy and representation, particularly in rural areas. The input datasets were combined without considering weighting as there is no guidance to indicate whether one factor is more important than another, and a sensitivity analysis was not carried out. Finally, how can you really validate the results coming out of such a top-down approach? Despite these limitations, Sanderson et al. (2002) are quick to point out that the map of human footprint is much too inexact for direct conservation purposes, and that what the map serves to do is illustrate the global picture of our current human footprint.

6.5 Conclusions

This chapter has outlined how data collected through the Geo-Wiki crowdsourcing tool has been used to create a map of human impact, which could be used as a spatial indicator of wilderness in a similar way to that undertaken by Sanderson et al. (2002) in identifying the 10 % wildest areas on Earth. As more crowdsourced data are collected in future Geo-Wiki campaigns and a denser network of points becomes available, we can continue to improve the crowdsourced map of human impact in the future.

This bottom-up approach represents a very different way of creating a map of human impact, which is normally derived through the top-down combination of various input datasets that reflect remoteness and/or biophysical naturalness such as the map of human footprint created by Sanderson et al. (2002). However, both bottom-up and top-down approaches have advantages and disadvantages. One way forward may be combining both approaches. For example, simple interpolation was used to illustrate how the map could be produced but there are better interpolation methods available that can use additional input data layers, e.g. road networks, population density, etc. to help guide the spatial allocation of human impact. Another approach would be to combine the global input data layers used by Sanderson et al. (2002) – although updated with the most current products available – in combination with the crowdsourced map of human impact in a type of multi-criteria hybrid approach. These types of approaches will be investigated as part of future work.

Acknowledgements This research was supported by the Austrian Research Funding Agency (FFG) via the LandSpotting (No. 828332) and FarmSupport (No. 833421) projects.

References

Applet, G., Thomson, J., & Wilbert, M. (2000). Indicators of wilderness: Using attributes of the land to assess the context of wilderness. *USDA forest service proceedings*, USDA, pp. 89–98.

Carver, S., Comber, A., McMorran, R., & Nutter, S. (2012). A GIS model for mapping spatial patterns and distribution of wild land in Scotland. *Landscape and Urban Planning, 104*, 395–409. doi:10.1016/j.landurbplan.2011.11.016.

Ellis, E. C., & Ramankutty, N. (2008). Putting people in the map: Anthropogenic biomes of the world. *Frontiers in Ecology and the Environment, 6*, 439–447. doi:10.1890/070062.

Fritz, S., See, L., McCallum, I., et al. (2011). Highlighting continued uncertainty in global land cover maps for the user community. *Environmental Research Letters, 6*, 044005. doi:10.1088/1748-9326/6/4/044005.

Fritz, S., McCallum, I., Schill, C., Perger, C., See, L., Schepaschenko, D., van der Velde, M., Kraxner, F., & Obersteiner, M. (2012). Geo-Wiki: An online platform for land cover validation and the improvement of global land cover. *Environmental Modelling and Software, 31*, 110–123.

Fritz, S., See, L., van der Velde, M., et al. (2013). Downgrading recent estimates of land available for biofuel production. *Environmental Science and Technology, 47*, 1688–1694. doi:10.1021/es303141h.

Gong, P., Wang, J., Yu, L., et al. (2013). Finer resolution observation and monitoring of global land cover: First mapping results with landsat TM and ETM+ data. *International Journal of Remote Sensing, 34*, 2607–2654. doi:10.1080/01431161.2012.748992.

Henry, D., & Husby, E. (1995). *Wilderness quality mapping in the Euro-Arctic Barents region: A potential management tool.* Northern wilderness areas: Ecology, sustainability (pp. 319–339). Rovaniemi: University of Lapland Arctic Centre Publications, University of Lapland, Arctic Centre.

Lesslie, R., Taylor, D., & Maslen, M. (1993). *National wilderness inventory. Handbook of principles, procedures and usage*. Canberra: Australian Heritage Commissions.

McCloskey, J. M., & Spalding, H. (1989). A reconnaissance level inventory of the amount of wilderness remaining in the world. *Ambio, 18*, 221–227.

Perger, C., Fritz, S., See, L., et al. (2012). A campaign to collect volunteered geographic information on land cover and human impact. In T. Jekel, A. Car, J. Strobl, & G. Griesebner (Eds.), *GI_forum 2012: Geovisualisation, society and learning* (pp. 83–91). Berlin/Offenbach: Herbert Wichmann Verlag.

Sanderson, E. W., Jaiteh, M., Levy, M. A., et al. (2002). The human footprint and the last of the wild. *BioScience, 52*, 891–904. doi:10.1641/0006-3568(2002)052[0891:THFATL]2.0.CO;2.

See, L., Comber, A., Salk, C., et al. (2013). Comparing the quality of crowdsourced data contributed by expert and non-experts. *PLoS ONE, 8*, e69958. doi:10.1371/journal.pone.0069958.

Theobald, D. M. (2004). Placing exurban land-use change in a human modification framework. *Frontiers in Ecology and the Environment, 2*, 139–144. doi:10.1890/1540-9295(2004)002[0139:PELCIA]2.0.CO;2.

Tobler, W. (1970). A computer movie simulating urban growth in the Detroit region. *Economic Geography, 46*, 234–240.

Chapter 7
Visualising Spaces of Global Inaccessibility

Benjamin D. Hennig

Abstract Wilderness is normally visualised using conventional mapping approaches showing the least touched spaces on the planet. Because most of the world's population lives in very limited spaces, this is a useful method of representation. The effects of human action, however, go much further than that, resulting in much of the land area being relatively close to humankind while the remotest and 'wildest' spaces are very little areas on a normal map. This chapter looks at alternative ways to visualise the most remote parts of the land area: by deploying a so-called gridded cartogram transformation to data about the (in)accessibility of a place, the resulting cartograms reveal the areas and the extent of the remotest spaces in a much less common way. Gridded cartograms are created by using an equally distributed grid onto which a density-equalising cartogram technique is applied. Each individual grid cell is resized according to its the average travel time to the nearest larger city. This technique is not only applied to the global scale, but also to regional and national-level data. The results are maps that give the remotest places most space and provide a unique and highly visual perspective on the spatial dimension of remoteness.

Keywords Cartograms • Gridded cartograms • Accessibility • Travel times • Remoteness

7.1 Introduction

Wilderness and remote areas are a diverse element in the patchwork of spaces that form the land surface of our planet. Only very small amounts of people are living in sparsely populated areas, which is an expression of the strong organisation of human societies to maximise those living in close relative proximity. More than half of the world's population now lives in areas categorised as cities, and although more than 95 % of the world's population live in approximately only 10 % of the land area, the

B.D. Hennig (✉)
University of Oxford, Oxford, UK
e-mail: benjamin.hennig@ouce.ox.ac.uk

S.J. Carver, S. Fritz (eds.), *Mapping Wilderness*,
DOI 10.1007/978-94-017-7399-7_7

remaining 90 % of the earth's land surface is far from being uniformly remote or even wild.

There are very different ways of how the un-built area that still makes the largest share of the land surface can be understood in terms of being under influence and in reach of human civilization. Only 15 % of people in rich countries live more than an hour of travel time from a city (of at least 50,000 people), while the same applies to 65 % of people living in the poor countries of the world. These are dimensions that conventional map projections hardly convey in their mere focus on generating images of the land surface and its physical extent. This chapter demonstrates a different cartographic approach to visualising and understanding these loneliest places on the planet by using a technique called a gridded cartogram transformation.

In this chapter, the gridded cartogram technique is used to visualise the relative distance of areas to the majority of people as calculated in an analysis of people's closeness (Uchida and Nelson 2010). The maps derived from the distorted grid show the physical space transformed according to the absolute travel time that is needed to reach the nearest major city by land transport averaged over the area of a grid cell. This results in a map that gives the remotest places the most space and provides a unique new and highly visual perspective on the spatial dimension of remoteness.

7.2 Data

The usefulness of gridded cartograms, like other cartograms, depends crucially on the data that goes into the cartogram transformation. Gridded cartograms require a different level of detail, but also a good variability within the quantitative data to result in meaningful and valuable visualisations.

A gridded population cartogram (Hennig 2013) is an effective visualisation of human space. It allows highlighting approximately 10 % of the land area where more than 95 % of the world's population live (Uchida and Nelson 2010), while the remaining 90 % of the land area vanishes in the visual display.

The optimal grid data for a gridded cartogram transformation thus needs a high spatial variation and a data distribution that has the highest values for the topic of interest in a limited amount of grid cells. For example, the major population densities cover approximately 10 % of the full grid (which is why a reverse view of the unpopulated areas appears much less striking). If that were not the case it would be less valuable, as it would not result in that unique new shape and could possibly be similarly well mapped with less methodological effort on a conventional map. One advantage of gridded cartograms is that of a suitable way to highlight selected areas that are in the centre of interest, while other spaces are blurred out.

This raises the question whether the least populated areas can be shown and emphasised with a similar approach. Defining wilderness and remoteness as merely unpopulated areas is less valuable for a cartogram visualisation, as most of the land surface of the planet still remains relatively unpopulated. A cartogram highlighting the least populated places (e.g. by reversing population data so that the least populated areas get the highest data values) results in a map that looks very similar to a conventional world map, with only some of the most populated regions being slightly smaller in size. Other concepts are therefore needed to look at the places that are least impacted by humankind.

The issue of less well-distributed data can be tackled if the underlying geographic dimension is more complex than a plain look at the depopulated areas. The interest in understanding less populated areas is also an issue of understanding the difference between the areas least and most touched by humanity, and the differences in their accessibility. Remoteness can therefore not only be a different way of understanding population patterns, but also of finding those areas that are still places of wilderness.

Some places are remoter than others regardless of their mere physical distance from people. One concept of understanding remote places in a more heterogeneous way is that of the accessibility and inaccessibility of places beyond their distance to major agglomerations. Data that analyses the (in)accessibility of places has been developed in a study on travel times to cities from any given point on the land surface by means of land transportation (Nelson 2008; Uchida and Nelson 2010). The data gives a different picture of remote places and helps to understand the accessibility of people and places beyond their relative distance in physical space. The data therefore also contains some information about the extent of natural areas that are less influenced by humankind.

The multifaceted dataset of travel times (Fig. 7.1) combines a number of different data sources from the physical and human environment to estimate the average travel times to the nearest major city defined by the authors of the study as one of the 8518 cities with 50,000 or more people. The data differentiates the concept of remoteness in a more detailed way by differentiating the least populated areas further than only making a statement about the population density. The indicators used

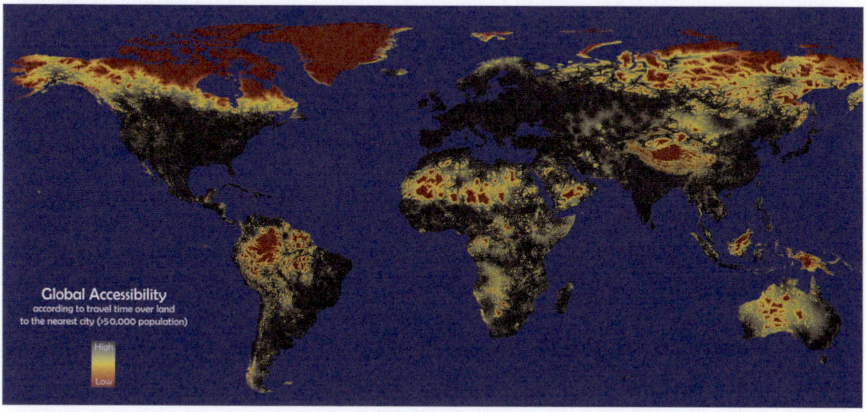

Fig. 7.1 Global accessibility according to travel time over land to the nearest city over 50,000 population (Own depiction by the author using data by Hengl 2011 and Nelson 2008)

to assess the travel times include all means of transport (rail and road network, rivers), travel times depending on the mode of transport and the kind of transport infrastructure, and the character and structure of the terrain (elevation, slope, land cover) (Nelson 2008; Uchida and Nelson 2010).

For the work presented in this chapter the information about the absolute travel time from a given point was converted into a grid with the least accessible places (and the longest travel time) containing the highest values. The data contained in the grid translates remoteness into a quantifiable measure that combines the human and physical space in one geospatial layer. The gridded dataset is then suitable for a gridded cartogram transformation based on the absolute travel time that is necessary to reach the nearest major city.

7.3 Methods

The maps presented in this chapter are based on research using gridded population cartograms as a novel approach to mapping quantitative data. The major improvement of the technique is the much higher precision in the transformed maps that allow cartogram-techniques to be used as an alternative map projection. This does not only result in maps that preserve the topology in the transformed spaces at a very high accuracy, but also in maps that allow other layers of geographic information to be reprojected accordingly (described in detail in Hennig 2013, 2014).

Gridded cartograms adapt a diffusion-based method for producing density equalising maps published by Gastner and Newman (2004). Their work outlines an advanced computer-generation of contiguous cartograms that Dorling (2006: 35) describes as "*one small step for two men, one giant leap for mapping*". These so-called diffusion cartograms transfer the physics of a linear diffusion process into the process of a map transformation.

The diffusion equation used in Gastner/Newman's approach emulates what happens when a liquid flows from higher to lower densities to smooth out the differences. In the map transformation, the diffusion equation smoothes out the differences in densities that are contained in the underlying data which results in distorted shapes of the original geographical areas. Like most cartogram approaches, Gastner/Newman's method is usually applied to countries or other larger administrative units. In a gridded cartogram however, the data is distributed onto an equally-sized grid or raster common to any GIS. This grid is then resized according to these quantities while preserving the relative position of each grid cell towards its neighbouring cells. The key unit in the cartogram transformation thus is not an arbitrarily or artificially defined administrative or other area, but a defined section of a map, with each areal unit having the same geographical extent (see more details described in Hennig 2013, 2014).

> Unlike conventional cartograms, gridded cartograms are capable of displaying additional layers of geographic information in large detail because they are based on a very accurate and neutral areal unit. Any way of putting information on a conventional map can be applied

to gridded cartograms. This can include any geographic subject with an allocation in the physical space (Hennig 2014).

The gridded cartogram approach results in a novel – yet unusual – map projection as the result are maps with a coherent and consistent base that never changes, while most other cartograms are more or less advanced (geo)graphical displays of quantitative data. Gridded cartograms preserve topological relations with a high spatial accuracy and allow other layers of spatial information to be re-projected accordingly.

In the original research on the technique, gridded population cartograms have proven to add further value when scaled to a particular area of interest, enhancing the level of detail and the information that can be derived from it considerably (Hennig 2013; Hennig et al 2010). Data availability defines the limitations of the technique, as the larger scales highly depend on the accuracy of the underlying raster of these maps.

A gridded population cartogram shows the world's geography in relation to the settlement patterns of humanity. This basic concept can be applied to any topic, meaning that it is not limited to socio-economic issues as they are normally chosen for cartogram depictions. On a global scale a gridded population cartogram was generated on the base of a ¼° population grid using data by CIESIN and CIAT (2005). Such a gridded population cartogram provided the basis for a first map transformation of the global accessibility.

The physical environment can be measured much more easily than the extent of the various dimensions of social space. Remote sensing technology and a huge network of measuring equipment constantly monitors a vast amount of data from the physical environment. This provides a solid base of quantitative data from the natural world that is of a very different level of detail than almost any other data that exists for the human world. Global accessibility combines these two worlds in an advanced way and brings data from the social and physical environment together.

Innovative geovisualisations generated from this data are rare, and most visualisations rely on the world map as the main choice for that data. Basic principles of cartographic practice are used to depict quantities. By applying the same principle of a grid that is transformed according to a quantitative indicator, a gridded cartogram transformation of global inaccessibility according to the total travel times to the nearest major cities can be realised.

The transformed grid thus shows each grid cell resized according to that absolute travel time that is needed to reach that area from the nearest major city by land transport. The method results in a map giving the remotest places most space on the map.

7.4 Results

Gridded population cartograms show land area distorted to the extent that every human being living on that surface is given the same amount of space on the map. In the resulting map every areal unit for which the total population is counted is transformed accordingly. Consequently the grid cell as the smallest areal unit in a

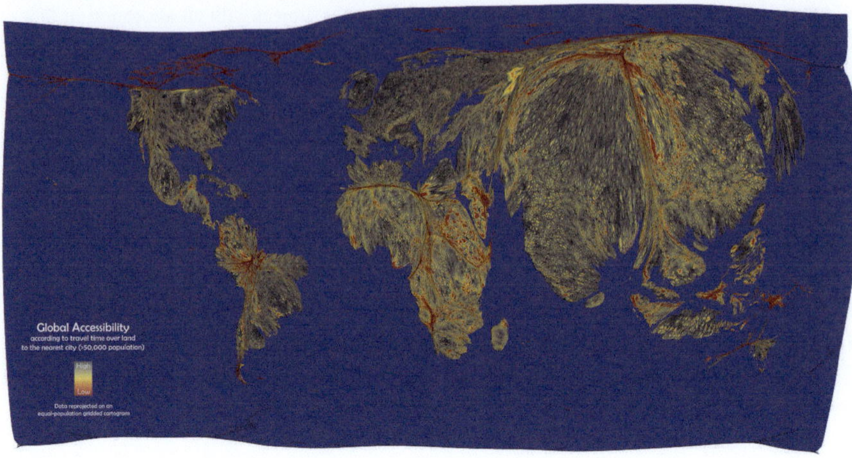

Fig. 7.2 Global accessibility according to travel time over land to the nearest city over 50,000 population shown on a gridded population cartogram (Own depiction by the author using data by CIESIN and CIAT 2005; Hengl 2011; Nelson 2008)

gridded cartogram is the smallest geographical reference to the physical space. The area that is covered in the original raster is the same area in the resulting gridded cartogram, hence the new maps show the number of people living in the same geographical space.

When transforming additional layers of geographic information alongside a gridded population cartogram, these topics can be interpreted accordingly and thus provide a deeper understanding of the topic in relation to the world population. In areas with few people, this additional layer is reduced in size because it does not relate directly to a larger amount of people. The underlying information is not removed in these areas of low population, but reduced in size and can be made visible by enlarging the area of interest. All original grid cells can be recognised in the gridded cartograms and no information is lost in the transformation process.

New layers of geographical information add further value to the gridded population cartograms. If seen as a new geographic map projection, then any ways of putting information on a conventional map can also work with gridded cartograms.

In the analysis of people's closeness, Nelson (2008) points out that only 15 % of people in rich countries live more than an hour of travel time from a major city, while the same applies to 65 % of people living in the poor countries of the world. The map showing travel time to major cities re-projected on a gridded population cartogram highlights this difference (Fig. 7.2). The colour scheme reflects the travel times to the nearest larger city as used in the conventional version of this map (Fig. 7.1). The value of the new projection becomes apparent, as the equal population projection demonstrates that not all parts of the world's population are equally accessible, and that in some regions large proportions of people live in relatively less accessible places. This is, for example, the case in some parts of Africa and the Arabian Peninsula where people live in relative remoteness to larger agglomerations

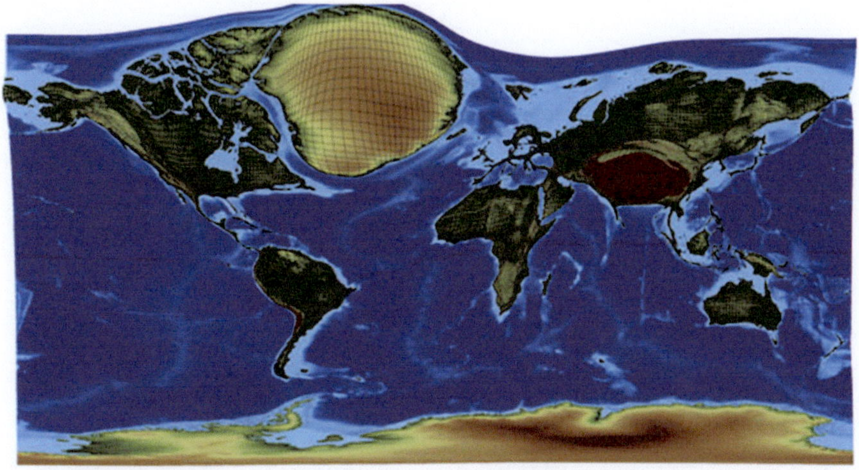

Fig. 7.3 Gridded cartogram of travel times to the nearest city over 50,000 population (Excluding Antarctica, own depiction by the author using data from CIESIN and CIAT 2005; Hengl 2011; Nelson 2008; NOAA 2009; USGS 2009)

although they are embedded in a denser population pattern, while this relative solitude hardly exists in Europe any more, where the transport infrastructure allows even the remotest parts that people see as distant and *wild to* be accessed much faster.

Population, however, is not the most suitable way of seeing and understanding the distribution of these remote areas. Here a gridded cartogram transformation of these spaces according to their relative inaccessibility creates a much more vivid impression of these last of the wild areas. Due to the nature of gridded cartograms, other geospatial layers can be transformed, so that for example the world's topography can be visualised in its correct geographic relation to the transformed maps. This can support the readability of the sometimes quite abstract shapes in the distorted maps.

The gridded cartogram of the remotest places on a global scale visualises the picture of a lonely planet (Fig. 7.3) where the spaces shown are those that are furthest away from those places of civilisation that define the twenty-first century. More than half of the world's population according to UN estimates now lives in cities (UNPD 2009), and this map shows those places that most of the people living in the world need the longest time to get to. It draws an image of the areas that are almost disconnected from those shrinking effects of globalisation, the real areas of wilderness that still surround us. This world map is the striking opposite representation of our image of a globalised and interconnected world, of those vanishing places that many people think do not exist any more.

The method of gridded cartograms is also scalable and can show remote places of a limited area, such as regions or individual countries. The resulting maps highlight further detail that cannot be seen on a global scale, which can give a new understanding of the places that are the valuable islands of remoteness within that

Fig. 7.4 Gridded cartogram of travel times to the nearest city over 50,000 population in Europe (Own depiction by the author using data from CIESIN and CIAT 2005; Hengl 2011; Nelson 2008; NOAA 2009; USGS 2009)

region. In some countries these are more, in others they are less remote, as the gridded country cartograms of travel times show.

The remotest places of Europe increase in size the further north one travels (Fig. 7.4). The southern regions mainly turn into remote areas where the terrain rises higher and the populated places dominate less of the landscape. Iceland is an island of remoteness in itself, with only its capital Reykjavík fulfilling the criteria for a major city. On the world map it is hard to see because much of the island is well connected only by a major ring road that circles around the island, but when changing the focus on Europe, it can be seen how the changing scope reveals different details and creates new understandings at different scales.

Even within the densely populated areas of Western Europe there are still some striking differences. The British Isles as part of the larger very populated regions in Europe show the dominating distances that increase the further north you come in Great Britain (Fig. 7.5). Furthermore, the relative remoteness of much of Wales can be seen, and the *Emerald Island* of Ireland shows up largely as a refuge of more natural environments where large parts of the island outside the major urban areas of Dublin and the slightly more densely populated northern region feature prominently in the transformed grid.

Europe's most populous country, Germany, shows a more even pattern with relatively remote rural areas and the urban centres being spread across the map (Fig. 7.6). It reflects the more even population distribution and is also a picture of a more balanced urban system in the country. The largest urban centres appear as nodes that

Fig. 7.5 Gridded cartogram of travel times to the nearest city over 50,000 population on the British Isles (Own depiction by the author using data from CIESIN and CIAT 2005; Hengl 2011; Nelson 2008; NOAA 2009; USGS 2009)

are often connected to each other via denser lines of less remote grid cells where grid lines seem to converge like a network of people across the country, with the most distant places in the country bulging out between this human cobweb.

The examples of gridded cartograms for remoter places show how different the emerging patterns become at various scales, and how well they create complementary understandings of the geography of the world, a continent, or a country.

Gridded cartograms can create a new image of the physical environment, which adds cartograms to the range of visualisations that can be used beyond the scope of merely socioeconomic issues. The resulting maps provide valuable new insights into the physical geography of the world and make quantitative data gathered from the environment more comprehensible without losing their geographical reference and accuracy. The visualisation created here can be useful for conveying quantitative geographical data in an understandable way to the public (e.g. shown in Holmes 2011), but may also be of further academic value as they can also provide new understandings of the patterns that the physical environment creates.

With the preservation of geographical references in the grid, a gridded cartogram is a novel type of geographic projection that is based on a replicable mathematical operation that turns the land area into spaces of equal quantities of any measurable

Fig. 7.6 Gridded cartogram of travel times to the nearest city over 50,000 population in Germany (Own depiction by the author using data from CIESIN and CIAT 2005; Hengl 2011; Nelson 2008; NOAA 2009; USGS 2009)

(or countable) topic. As presented here, gridded cartograms provide a serious alternative to traditional mapping methods.

7.5 Discussion

Global spaces of remoteness and inaccessibility are the last refuges of wilderness that are not untouched by humankind, but that are the precious places where we preserve some of what the world was like before we started making it the human planet (Steffen et al. 2011). While these spaces are in decline, the most remote areas are only a fraction of our land surface that we see in conventional maps. Cartograms can provide a powerful visualisation tool that provides the base to redraw the geography of the planet in diverse ways and to highlight these diverse natures of the world.

The constraint of arbitrary administrative boundaries that usually build the base for a cartogram transformation does not exist anymore in the gridded cartograms. The high resolution and good quality of the now existing global datasets of a wider range of issues related to human activity provides a solid base for this geographic transformation.

Visualising spaces of wilderness as demonstrated with the concept of global accessibility can be achieved in manifold ways. Beyond the conventional mapping approaches that are normally used, the use of a gridded world population cartogram as a basemap highlights the relative accessibility of people on the planet. It demonstrates that some parts of the world's population are less accessible than others and which matters in a world that increasingly depends on global links and interrelations.

The human space in the centre of the map projection shows the dominating spatial element of what has recently often been described as the Anthropocene. But this is only one way of transforming space and looking at the dimensions that shape our planet. The general methodology can be extended even further and provide novel ways of visualising any quantitative geospatial data in cartographic ways.

Global accessibility itself can be converted to a quantitative measure that can be applied to the grid cells as the basis for a cartogram transformation, as demonstrated here with the least accessible spaces. After the actual transformation, these spaces reflect the inaccessibility, as longer travel times to get to an area result in an increased the size of the corresponding grid cell. Conventional map projections limit the range of visualisation methods that are suitable to show quantitative data. In a plane two-dimensional map, this can be done in choropleth form with distinct colour schemes (Fig. 7.1), but these have the disadvantage to require a rather abstract translation of the colours into a third dimension of quantities.

3D visualisations provide an alternative, but therefore have to be interactive or require a number of several individual perspectives to be able to understand the third dimension in a two dimensional form (on paper or on screen). The use of graphs and charts in contrast loses the advantage of a map as a direct geographical reference.

The gridded cartogram technique can provide an alternative solution to this problem that retains a very high geographic accuracy in the grid but makes the quantitative dimension visible in the changing sizes of the grid cells. The emerging patterns retain the geographical reference, but can become more abstract. The more unusual patterns, however, create a unique vivid appearance of the displayed topic and open a new dimension of understanding the large amount of underlying data.

Gridded cartograms created from geospatial data are not a single new map projection, but a technique that makes geographically relevant topics from the physical and social world come alive and thus represent a multitude of unique map projections. The gridded cartograms have the potential to create a different understanding of the real extent and impact of a geographical topic. They may be more unusual and still unconventional, but provide alternative ways of geovisualisation while retaining core cartographic elements of spatial representation.

Crucial for the further application of gridded cartograms are the underlying datasets. Data for the gridded cartograms must not have any gaps but have to be complete over the entire grid covering the area of interest. Only then can one create a coherent and valid transformation that does not result in a misleading representation. That is crucial for any cartogram, as gaps in the data cannot be dismissed like in a conventional map. Therefore the creation of gridded cartograms requires a high effort, which is often greater than that of other cartograms where only a limited number of data values is needed for the transformation.

With huge advances in data availability in recent years, and with growing demand for gridded data in scientific modelling, these datasets have become increasingly available and do not always need to be created. Therefore, this alternative way of visualising them provides a great opportunity to enhance the understanding of these complex datasets in the different scientific domains.

For research on wilderness and remote areas, global accessibility therefore is only one concept that provides a suitable quantitative dataset that can be applied to a gridded cartogram transformation. Wild spaces have a multitude of expressions. (In)accessibility is an issue that is closely linked to human activity, while some other areas of research may be less interested in the accessibility, but in other indicators that define a wilderness area. The range of potential applications therefore is large and mainly requires a creative and unconventional approach to data preparation.

The gridded cartogram technique is still a time-consuming approach that requires a lot of effort in pre-processing and transforming the data. Tools that simplify the methodological effort will have to be developed to make this process easier. Processing times are another crucial issue. But with increasing computing capabilities and with more data resources becoming available, this is likely to become more straight-forward in the forthcoming years. A clear objective must be the simplification of the method to implement it in common workflows of visualising geospatial data.

The concept of creating new gridded data beyond mono-thematic quantitative data by combining data sets, like the calculation of travel times from a range of indicators of the physical and social environment, demonstrates the benefit that new unconventional approaches can have. Much scientific effort is put into the creation of indices and similar models that often also bridge the gap between human and physical geography. Global accessibility does not look at either one or the other side but aimed at getting understandings of the different interrelations and connections and see how they create new patterns and shapes of humankind. But why stop there and not think further also for the visualisation of such data: Wilderness areas are the last spaces of the wild. Why not rediscover these hidden gems by looking at them in new ways?

7.6 Conclusion and Outlook

This chapter presented a visualisation approach using cartograms as a way to visualise spaces of global inaccessibility. The result are maps of the remotest parts of the planet that add that dimension of (in)accessibility visually to the map while preserving some key features that characterise conventional map projections. The results are not only geographically accurate depictions of the underlying data, but are also scalable to show global, regional and national variation within the data.

Adapting the famous expression commonly used in conjunction with pictures, the examples outlined in this chapter show that *a map is worth a thousand words*. Techniques of geographic visualisation can be a useful tool to translate complex ideas into thought-provoking images where the traditional idea of a map does not always have to be the concept for a spatial data visualisation.

The maps presented here are a novel approach of using cartogram transformations for the visualisation of quantitative data not only related to the social dimensions, but also with a direct relation to the diverse physical environments that shape our perception of the planet. The concept may not be useful for all applications and studies of wilderness, but it extends the range of techniques that we can use to provide different perspectives on the remotest spaces of the planet.

Gridded cartograms mark a new way of utilising cartogram creating techniques at various scales. The geographic reference to the physical space is kept in the cartographic transformation, allowing other information to be transformed alongside the defining indicator of the actual cartogram. The examples in this chapter demonstrated that this unique characteristic of gridded cartograms is a key element to enhance the value of the analytical and visualisation capabilities of cartograms and make them similar to an unusual but powerful map projection that gives us a different understanding of the relation between the cartogram and any other layers that are mapped with it.

The range of potential application for these maps is only limited by data availability and thematic areas of interest – and one's imagination. Instead of using topographic information as in the maps presented in this chapter, possible uses could highlight topics ranging from biodiversity, environmental risk factors to human activity in these remotest places outside the direct range of humanity. While these maps show the most inaccessible spaces on the land surface, such visualisations can be one way of highlighting the threats affecting even these remote areas. Where performed accurately, gridded cartograms have the capability to function like a magnifying glass over the key area of interest, which in many cases is omitted in conventional map projections, while at the same time preserving the geographic accuracy of the information that they contain.

No map projection is perfect. Sometimes it can be helpful to be more creative and think outside the box of cartographic conventions that we have learned to accept as normal over the decades and centuries. Today's world is different than that of the times when many of our cartographic techniques were developed. We live on an ever-changing planet where the interrelation between humans and their environment has become one key element that shapes it. To understand these changes, we have to change our perspectives and to find new ways of understanding these complex spaces.

Gridded cartograms are not perfect either. First and foremost they are still time-consuming pieces of work that require a lot of effort in their production. Once the underlying data has been obtained or generated, the real work of adjusting and transforming it starts. Putting these technical obstacles aside, they are unusual yet striking representations of the spaces that are hardest to get to. They provide one more way of seeing the world beyond the box of maps we normally use.

References

CIESIN and CIAT (Center for International Earth Science Information Network/Centro Internacional de Agricultura Tropical, Socioeconomic Data and Applications Center (SEDAC), Columbia University, NY). (2005). *Gridded Population of the World Version 3* (GPWv3): Population grids. http://sedac.ciesin.columbia.edu/gpw/global.jsp. Last accessed 1 May 2013.

Dorling, D. (2006). New maps of the world, its people, and their lives. *Society of Cartographers Bulletin, 39*(1–2), 35–40.

Gastner, M. T., & Newman, M. E. J. (2004). Diffusion-based method for producing density equalizing maps. *Proceedings of the National Academy of Sciences of the United States of America, 101,* 7499–7504. doi: arXiv:physics/0401102v1 [physics.data-an].

Hengl, T. (2011). *Global datasets*. http://spatial-analyst.net/wiki/index.php?title=Global_datasets. Last accessed 1 June 2011.

Hennig, B. D. (2013). *Rediscovering the world: Map transformations of human and physical space*. Berlin/Heidelberg: Springer. doi:10.1007/978-3-642-34848-8.

Hennig, B. D. (2014). Gridded cartograms as a method for visualising earthquake risk at the global scale. *Journal of Maps, 10*(2). doi:10.1080/17445647.2013.806229.

Hennig, B. D., Pritchard, J., Ramsden, M., & Dorling, D. (2010). Remapping the world's population. Visualizing data using cartograms. *ArcUser, 2010*(1), 66–69.

Holmes, N. (Ed.). (2011). *How to land a jumbo jet, a visual exploration of travel facts, figures and ephemera*. Victoria: Lonely Planet.

Nelson, A. (2008). *Travel time to major cities*. Report. Luxembourg: European Commission.

NOAA (National Oceanic and Atmospheric Administration). (2009). *ETOPO1 global relief model*. http://www.ngdc.noaa.gov/mgg/global/global.html. Last accessed 1 May 2013.

Steffen, W., et al. (2011). The Anthropocene: From global change to planetary stewardship. *AMBIO, 40*(7), 739–761.

Uchida, H., & Nelson, A. (2010). Agglomeration index: Towards a new measure of urban concentration. In J. Beall, B. Guha-Khasnobis, & R. Kanbur (Eds.), *Urbanization and development: Multidisciplinary perspectives* (pp. 41–60). New York: Oxford University Press.

UNPD (United Nations Population Division). (2009). *World urbanization prospects: The 2009 revision* (Report). New York: United Nations Publications.

USGS (US Geological Survey). (2009). *GTOPO30*. http://eros.usgs.gov/#/Find_Data/Products_and_Data_Available/gtopo30_info. Last accessed 1 June 2011.

Chapter 8
Addressing Weak Legal Protection of Wilderness: Deliberate Choices and Drawing Lines on the Map

Kees Bastmeijer

Abstract Wilderness areas are characterized by a relatively high degree of naturalness, the absence of proof of modern human society (e.g. roads, buildings, bridges, motorized transportation) and a relatively large size (Dudley, N. (Ed.). (2008). *Guidelines for applying protected area management categories*. Gland: IUCN. Available at: http://data.iucn.org/dbtw-wpd/edocs/2008-028.pdf). Worldwide, wilderness areas are becoming more scarce and this chapter focuses on the role of law in protecting such areas. The discussion starts with an analysis of the historic human-nature attitude in Western society and how this attitude has influenced legal concepts regarding private property on land and territorial sovereignty. It will be shown that these concepts have stimulated active land transformation by humankind and that (as a consequence) wilderness protection is not embedded in our Western legal roots. Next, the discussion focuses on the response to the increasing awareness of the downside of modern human civilization: a changing human-nature attitude in the Nineteenth Century and the adoption of a large number of international nature protection conventions in the Twentieth Century. However, all this 'law making' has not resulted in comprehensive wilderness protection at the global or regional level, which may be explained by a number of important weaknesses in these conventions and their implementation. Probably, many of these weaknesses have much to do with weaknesses of humankind itself, such as the difficulty to accept limitations to our social and economic ambitions and our disability to deal with accumulative impacts. Against the background of these discussions, the final part of this chapter discusses options for strengthening wilderness protection with an emphasis on the importance of making deliberate policy choices to protect wilderness.

Keywords Wilderness • Civilisation • Exploitation • Private property • Romantic period • Nature protection conventions • European Union • Protected areas •

This chapter builds on Kees Bastmeijer, Introduction: An international history of wilderness protection and central aim of this book. In: *Wilderness Protection in Europe. The Role of International, European and National Law*, Cambridge University Press, 2016.

K. Bastmeijer (✉)
Tilburg Law School, Tilburg University, Tilburg, The Netherlands
e-mail: c.j.bastmeijer@tilburguniversity.edu

S.J. Carver, S. Fritz (eds.), *Mapping Wilderness*,
DOI 10.1007/978-94-017-7399-7_8

Sustainability • Biodiversity conservation • Non-use • Wilderness protection • Accumulative impacts • Politics

8.1 Introduction

Twenty-five years ago, Philip Dearden stated that "[u]p to the late twentieth century, wilderness has been, by and large, a by-product" as it is "what has been left after the 'good' land has been taken for agriculture, forestry, mining, urbanization, industry and every other conceivable land-use" (Dearden 1989, p. 206). In most parts of the world, this has not changed much. Steve Carver explains that recent mapping work has shown patterns of wilderness to be strongly influenced by latitudinal and altitudinal gradients that place physical limits on agriculture and forestry as well as cultural and political gradients that place limits on human land use (Carver 2016). This might easily be explained by the fact that many forms of modern human land use have a strong tendency to affect one or more of the main qualities that characterize wilderness areas (Dudley 2008; Kuiters et al. 2012; Wild Europe Initiative 2013): (a) naturalness (native species and ecosystems and free functioning natural processes), (b) the absence of (and minimum distance from) roads, buildings, bridges, tracks, cables or other proof of modern human society, and (c) the relatively large size of the area. However, from a legal perspective, one might consider it quite remarkable that wilderness areas are indeed just 'left-overs' of human caused land transformation. Has the law not been able to prevent wilderness loss, and if this is true, how may this be explained? This is the major topic of this chapter.

The discussion starts with an analysis of the historic human-nature attitude in Western society and how – during the last centuries – this attitude has influenced legal concepts regarding private land property and territorial claims by states (Sect. 8.2). It will be shown that these concepts constitute stimuli for land transformation and that wilderness protection is not embedded in our Western legal roots. Next, the discussion focuses on the response to the increasing awareness of the downside of modern human civilization: a changing human-nature attitude in the nineteenth century and the adoption of a large number of international nature protection conventions in the twentieth century (Sect. 8.3). This discussion is followed by an identification of some important weaknesses of these conventions and their implementation to explain why intensive law making has not resulted in comprehensive wilderness protection at the global or regional level (Sect. 8.4). In the final part of this chapter, options for strengthening wilderness protection are discussed, with special attention for the importance of 'drawing lines on maps' and deliberate choices to protect wilderness (Sect. 8.5).

8.2 Tensions between Western Legal Roots and Wilderness Protection

In the Western world humankind has for centuries taken a dominant position over nature. Explanations for this attitude are diverse. Some go far back in time and claim that the split between people and nature coincided with the origin of agriculture (Wells 2010) and the domestication of animals (roughly around 8000–4000 BC, depending on the geographical region) (DeMello 2012, chapter 4). Others refer to the ancient Greeks (Zweers 1995, p. 27–28; Passmore 1980). For instance, Paul Cliteur refers to a phrase in Aristotle's *Politica* (Cliteur 2005), where Aristotle states, in chapter VIII of the first book:

> It is evident that we may conclude of those things that are, that plants are created for the sake of animals, and animals for the sake of men; the tame for our use and provision; the wild, at least the greater part, for our provision also, or for some other advantageous purpose, as furnishing us with clothes, and the like (Aristotle Pol.).

Lynn White Jr. has stated in his much-debated *Science*-article (White 1967), that the Judeo-Christian tradition constitutes the main source of the dominant attitude of humans over nature (Minteer and Manning 2005). One could also refer to views expressed by lawyers and philosophers in the sixteenth and seventeenth centuries. For instance, Hugo Grotius states in his *Mare Liberum* (The Freedom of the Seas) that God created nature for mankind (Grotius 1609, p. 22). He believed that "God had not given all things to this individual or to that, but to the entire human race" (Grotius 1609, p. 24). John Locke (1632–1704) shared this view: "The earth and all that is therein is given to men for the support and comfort of their being" (Locke 1690, chapter 5, par. 25; Snyder 2007, p. 15). A possibly even stronger source of human dominance over nature is the 'mechanization of nature' in the theory of Descartes (1596–1650) (Verbeek 2007, p. 37).

Although in all these time periods contra-arguments have also been expressed (e.g. by Spinoza (1632–1677), the dominant view has been that nature was meant to serve humankind. This also implied that humans had the right to transform nature for their own benefit. According to Locke, the fact that nature was meant to benefit humankind constituted the fundament for acquiring components of nature as private property:

> The earth and all that is therein is given to men for the support and comfort of their being. And though all the fruits it naturally produces, and beasts it feeds, belong to mankind in common, as they are produced by the spontaneous hand of nature, and nobody has originally a private dominion exclusive of the rest of mankind in any of them, as they are thus in their natural stage, yet being given for the use of men, there must of necessity be a means appropriate them some way or other before they can be of any use, or at all beneficial, to any particular men (Locke 1690, chapter 5, par. 25; Snyder 2007, p. 15).

Before Locke, also Hugo Grotius had expressed the view that nature itself was the source for this extension of appropriation: food and drinks implicate a form of ownership because consumption is exclusive; it cannot be consumed by someone else at the same time, and according to Grotius this had constituted the fundament for a process of appropriation of nature into private property: Starting with food and drinks, followed by *"things of the second category, such as clothes and movables and some living things"* the subjects of private ownership were extended (Grotius 1609, p. 24). According to Grotius also the land would be divided into property (Schrijver and Prislan 2009, p. 172):

> When that had come about, not even immovables, such, for instance, as fields, could remain unapportioned. For although their use does not consist merely in consumption, nevertheless it is bound up with subsequent consumption, as fields and plants are used to get food, and pastures to get clothing. There is, however, not enough fixed property to satisfy the use of everybody indiscriminately (Grotius 1609, p. 24–25).

Thus, nature was meant for mankind and could therefore be appropriated as private property to enable human use. The reverse was also true: in this time period (1600–1900) appropriation of nature (e.g., land) was only justified if the land would actually be exploited. This opinion may be found in the works of John Locke as well as in legal visions in the last centuries on the criteria for legal land claims under international law: a nation is not allowed to appropriate more land than it can populate, cultivate and govern. Thomas Willing Balch discussed this in detail in 1910 and quoted several jurists of the eighteenth and nineteenth century (Balch 1910), including the Swiss jurist De Vattel (*Le Droit des gens*, 1758; English translation 1867):

> But it is questioned whether a nation can, by the bare act of taking possession, appropriate to itself countries which it does not really occupy, and thus engross a much greater extent of territory than it is able to people or cultivate. It is not difficult to determine that such a pretension would be an absolute infringement of the natural rights of men, and repugnant to the views of nature, which, having destined the whole earth to supply the wants of mankind in general, gives no nation a right to appropriate to itself a country, except for the purpose of making use of it, and not of hindering others from deriving advantage from it. (De Vattel 1758, p. 98–99)

It is clear that this process of appropriation of nature as private property and the claims of new land as state territory has resulted in a continuing process of cultivation of nature. As explained by George P. Marsh in 1867 in the preface of his famous "Man and Nature':

> The extension of agricultural and pastoral industry involves an enlargement of the sphere of man's domain, by encroachment upon the forests which once covered the greater part of the earth's surface otherwise adapted to his occupation. [...]. Lands won from the woods must be both drained and irrigated; river banks and maritime coasts must be secured by means of artificial bulwarks against inundation by inland and by ocean floods; and the needs of commerce require the improvement of natural, and the construction of artificial channels of navigation (Marsh 1867).

The above discussion illustrates the strong belief in past centuries that the earth is meant for humankind and that the development of private property and territorial claims of states are only justified if the land would actually be exploited for the benefit of mankind. From the perspective of legal protection of wilderness, this is relevant as these strong roots in Western legal thinking were stimuli for transformation of the earth's surface and made wilderness, as defined and valued today, not logical or even problematic. As will be discussed below, attitudes have changed; however, these legal roots may still have their explicit or implicit influences and may still constitute a hurdle in legal protection of wilderness.

8.3 Downside of Civilisation and the Legal Response: The Development of International Nature Protection Conventions

Probably based on the above historic roots in combination with other factors, particularly human population growth, humankind in the Western world has proven to be very successful in using the natural resources of the earth to ensure plenty of food and materialistic wealth for at least a large part of the population. As explained by Crispin Tickell *"unlike other animals, we made a jump from being successful to being a runaway success [...] because of our ability to adapt environments for our own uses in ways that no other animal can match"* (Tickell 1993, p. 219; Roberts 1996). However, particularly during the nineteenth century people in Western societies became increasingly aware of the downside of these developments. In scientific and more popular literature these views were refined: *"Doubts and hesitations had arisen about man's place in nature and his relationship to other species. [...] A closer sense of affinity with the animal creation had weakened old assumptions about human uniqueness"* (Thomas 1983, p. 243). In this context, the works of Charles Darwin (1809–1882) (Darwin 1871) and his contemporaries such as Thomas Henry Huxley (1825–1895) (Huxley 1863) are of great importance (Cliteur 2001, p. 6). An important factor for increased appreciation for nature in this period was also the downside of life in the city: increasing air pollution, crime and diseases. Keith Thomas offers many splendid quotations and sources that show that the idealization of cities of earlier times had to give way to an increasing appreciation for country life and nature (Thomas 1983, p. 242–254).

However, increased appreciation of nature in the period of romanticism was not yet reflected in law making. Certainly, the downside of our own success in the form of over-exploitation of wild species of plants and animals was recognized, but resulted only in treaties that aimed at the protection of plants and animals that were useful to mankind. Examples of early treaties are the 'Convention for the Protection of Wild Animals, Birds and Fish in Africa' (Sands 1995, p. 27), signed in 1900 by the colonial authorities of Africa in London, and the European Convention to Protect Birds Useful to Agriculture of Paris, 1902 (Bowland 1989). As Bowland notes: *"they were concerned predominantly with direct and immediate human interests rather than motivated by any more elevated or altruistic ideals [and] they encouraged the destruction of certain creatures that were judged harmful to those interests"* (Bowland 1989, p. 487; Van Heijnsbergen 1997, p. 130. For other examples, see the Convention between the U.S. and Great Britain (for Canada), 16 August 1916 (39 Stat. 1702). The U.S. entered into similar agreements with Mexico (1936), Japan (1972) and the Soviet Union (1976). The United States already had an Act to protect birds: the Lacey Act (16 U.S.C. § 701, 25 May 1900)).

Changing appreciations of nature also applied more specifically to relatively untouched natural areas, particularly in North America: *"Wilderness had once been the antithesis of all that was orderly and good—it had been the darkness, one might say, on the far side of the garden wall—and yet now it was frequently likened to*

Eden itself" (Cronon 1995; Nash 2001). The strong advocacy by Thoreau, Muir and others to set aside untouched nature in North America is well known, although it would be a mistake to think that it was their intention to separate humans from nature: *"Though Muir like other romantics denied that the earth was made for man, it was for men's spiritual salvation that they sought to save wild nature [...]"* (Lowenthal 2000). Also in Western Europe the special value of wilderness was emphasized in literature and other works of art as well as in legal and policy debates (Fisher 2016). For instance, in the Netherlands H.P Gorter, with reference to Frederik Willem van Eeden Sr. (1829–1901), explained in relation to the second half of the nineteenth century:

> It was at that time, that they, who looked further into the future, saw the signs that the wilderness, which at one time covered our land as far as the eye could see, would become a scarcity, and that it would become necessary to defend the 'right of the wilderness' (Gorter 1956, p. 11).

Only in North America, this development resulted in an international convention with explicit attention for wilderness protection. According to Article II of the 'Convention on Nature Protection and Wild Life Preservation in the Western Hemisphere', that was adopted in 1940, *"[t]he Contracting Governments will explore at once the possibility of establishing in their territories national parks, national reserves, nature monuments, and strict wilderness reserves as defined in the preceding article."* Art. 1(4) defines 'strict wilderness reserves' as: *"A region under public control characterized by primitive conditions of flora, fauna, transportation and habitation wherein there is no provision for the passage of motorized transportation and all commercial developments are excluded."* Article IV states that "[t]he Contracting Governments agree to maintain the strict wilderness reserves inviolate, as far as practicable, except for duly authorized scientific investigations or government inspection, or such uses as are consistent with the purposes for which the area was established."

After World War II, attention for the negative 'externalities' of human exploitation at the global level intensified. The foundation of the World Conservation Union (IUCN) in 1948 and the organisation of the UN 'Conference on the Conservation and Utilisation of Resources' in 1947 (17 August–6 September 1949, Lake Success, New York) are among the important events of the previous century (Sands 1995, p. 31–32; Goodrich 1951). Nico Schrijver states that during the latter conference concerns were raised with regard to the irresponsible exploitation of natural resources, but then refers to a passage in the report of the conference that shows that most experts thought that with the right use of technologies and the prevention of squandering it should be possible to offer a higher standard of life to a bigger world population (Schrijver 2008, p. 37–38). Scarcity should be dealt with partially by technological developments. This confidence in technological solutions was strong in this time period:

> The past was blamed; the present was smarter. [...] With few exceptions optimism prevailed. Many showed extreme complacency in the face of threats that now seem evident. Humans were thought incapable of significantly changing global climate; nuclear-fission

> wastes were wholly benign; wise management would rebuild impoverished soils. Why worry about nuclear by-products; past fears of technology had always come to naught. In sum, environmental impacts scared only scientific idiots and crackpots (Lowenthal 2000, p. 8).

However, during the second part of the twentieth century *"environmental impacts are increasingly seen as global and interrelated, complex and unknowable, long-lasting and perhaps irreversible"* (Lowenthal 2000, p. 10). In 1969, the American National Research Council emphasized that humankind is in an extreme time period, characterized by a disbalance between development and available natural resources:

> It now appears that the period of rapid population and industrial growth that has prevailed during the last few centuries, instead of being the normal order of things and capable of continuance into the indefinite future, is actually one of the most abnormal phases of human history (King Hubbert 1969, p. 238).

A few years later, at the occasion of the United Nations Stockholm Conference on the Human Environment (1972), the international community explicitly stressed the need for fundamental change:

> A point has been reached in history when we must shape our actions throughout the world with a more prudent care for their environmental consequences. Through ignorance or indifference we can do massive and irreversible harm to the earthly environment on which our life and wellbeing depend (Stockholm declaration 1972, preamble, para. 6).

This acknowledgement has resulted in several decades of law-making in respect of many environmental concerns. In relation to the natural world, many international conventions, regional binding instruments and domestic laws have been adopted, particularly since the early 1970s, to protect the variety of life forms (species of plants and animals), habitats and ecosystems (biological diversity, hereinafter: biodiversity) (Birnie et al. 2009, p. 588). Important conventions include the Convention on Wetlands of International Importance (Ramsar Convention), the Convention on the Conservation of Migratory Species of Wild Animals (Bonn Convention), the Convention on the Conservation of European Wildlife and Natural Habitats (Bern Convention), and the Convention on Biological Diversity (CBD). As most of these conventions have been signed and ratified by more than 175 states, this international nature protection law may be considered as 'global law'. In addition, much nature protection law has been developed at the regional and domestic level. In the European Union the Bird Directive and Habitat Directive constitute a system of fairly strict protection of wild species, habitats and important natural sites (Natura 2000 sites). And even for 'far away-places' such as Antarctica much law has been developed to protect the natural environment. For instance, with the Protocol on Environmental Protection to the Antarctic Treaty, the whole region south of 60° south latitude has been designated as "natural reserve, devoted to peace and science" (Antartic Treaty protocol 1991, art. 2).

8.4 The Role of Law in Protecting Wilderness: Some Fundamental Weaknesses

It is correct to say that *"[m]ost likely the global situation state of wetlands, properties of outstanding universal value, endangered species of wild fauna and flora, migratory species of wild animals and biodiversity as such would have been considerably worse without the existence of the conventions"* (Koester 2012, p. 70); however, the great number of monitoring reports and effectiveness studies obliges us to acknowledge the limited effectiveness of these agreements. *"It is well established that losses in biodiversity are occurring globally at all levels, from ecosystems through species, population, and genes"* (World Resources Institute 2005, p. 834; CBD 2010, p. 9), and *"[t]he five principal pressures directly driving biodiversity loss (habitat change, overexploitation, pollution, invasive alien species and climate change) are either constant or increasing in intensity"* (CBD 2010, executive summary).

These and other causes have also resulted in the current situation in which only about 30 % of the earth's land surface can still be qualified as relatively untouched by humans ('wilderness') (Kormos and Locke 2008). The causes for this limited effectiveness of nature conservation law are multiple (Baakman 2011; Caddell 2005; Morgera and Tsioumani 2011); however, four more fundamental characteristic weaknesses may be identified that make legal wilderness protection particularly problematic.

8.4.1 Strong Focus on Sustainability and Biodiversity Conservation

From the perspective of protecting wilderness, an essential weakness of most international nature conservation conventions is that these legal instruments focus strongly on two main objectives: 'sustainable development' and the protection of 'biological diversity' (biodiversity). Both concepts are in part interrelated and appear to focus on establishing an acceptable balance between economic, social and ecological needs (IUCN 2004, p. 5). The discussion under the CBD emphasizes this, and also other international nature protection agreements are built upon this 'balancing of interest approach'. For instance, a central instrument of the Wetland Convention (Ramsar Convention) is the obligation to ensure the 'wise use' of important wetlands.

The notion of balancing of interests is inherent to sustainable development, but often it is not fully clear what natural values or interests should be taken into account. In view of the CBD and other nature protection conventions, the conservation of the world's diversity of plants and animals, habitats and ecosystems clearly is a central aim of international nature conservation law; however, it is unclear to what extent these conventions also aim to protect relatively undisturbed wilderness areas, their characteristic qualities and related values. The strong emphasis on biodiversity conservation may have resulted in narrowing the scope of legal nature protection:

> The greatest push for conserving protected areas has come with the recognition that biodiversity is also crucial for human survival. [...] Along with this concept came a change in the view of what should comprise a protected area. Instead of untouched wilderness, current protected areas are frequently made up of areas of supervised human activity (Jeffery 2004, p. 14).

This strong focus on biodiversity conservation has also resulted in an approach that is characterized by protecting 'special' species, habitats and ecosystems only (Doremus 1999). And being 'special' is generally not a good thing as it often means 'being threatened'. The legal protection in the early conventions mentioned above and in domestic laws was particularly attributed to those species that were almost extinct (Birnie et al. 2009). In the literature this type of nature conservation is referred to as 'deathbed conservation' (Trouwborst 2008). The more modern conventions clearly have a broader purpose and scope; however, many of the more strict prohibitions and requirements in the conventions apply only to 'special' species and particular types of ecosystems that have been listed (Bonn Convention, Bern Convention, Ramsar Convention). On the one hand this approach is logical as particularly threatened natural values require a priority in protection and there are quite a number of success stories of such focused approaches (Deinet et al. 2013). On the other hand however, this strong biodiversity approach in combination with the focus on 'special values' may constitute a weakness in legal protection of areas with high wilderness qualities: legal protection of such wilderness areas depends heavily on the question whether the areas meet the specific criteria for designation and protection under the relevant legal regime.

For instance, important natural areas in the European Union must be designated and protected as Natura 2000-sites; however, areas only qualify if they are the "most suitable territories in number and size" (article 4(1) Bird Directive) for birds listed in Annex I of the Bird Directive or if they host "natural habitat types" listed in Annex I and/or "species" listed in Annex II of the Habitat Directive (article 4(1) Habitat Directive). Consequently, wilderness areas that have no relevance for such listed species and habitat types do not have to be designated and protected under this EU system. Research on mapping wilderness in respect of the extent of overlap between wilderness qualities in Europe and Natura 2000-status is ongoing, but the currently available knowledge indicates that in Europe "many areas of *de facto* wilderness are still going without protection" (Carver 2016).

Furthermore, if wilderness areas do also qualify as a Natura 2000-site and have been designated accordingly, protection of the wilderness qualities of these areas is not automatically guaranteed. Article 6 of the Habitat Directive contains a strict regime of protection; however, the aim of this regime is to prevent significant effects on the site "in view of the site's conservation objectives." As these objectives are likely to focus on the habitat types and species for which the site has been designated, adverse effects on the wilderness qualities of the site may only be considered as 'significant' if these effects are also adverse for the conservation or recovery of the relevant habitat types and species. As stated by the Advocate General of the Court of Justice of the EU in the Waddensea case, *"[a]dverse effects, which are not obvious in view of the site's conservation objectives, may be disregarded"* (Case C-127/02, Opinion AG Kokott, para. 72). For instance, establishing permanent infrastructure in a Natura 2000-site will limit the wilderness qualities of the site, but may not be considered significant under the Natura 2000-regime if such infrastructure is not causing negative effects in view of the conservation objectives of the site. As explained by Advocate General Mazak in a case relating to wind turbines: *"the referring court, the applicant companies and the Commission correctly state that the classification of a zone as a site of Community importance or special protection area forming part of the ecological network Natura 2000 does not result in all construction therein being banned in accordance with the Birds and Habitats Directives"* (Case C-2/10, opinion AG Mazak, para. 30). Another example relates to the upgrade of a road in lynx habitat in Spain: the road was fenced over 9300 m on both sides of the road and although this project limits the wilderness qualities of the site, the Court concluded that significant effects for the lynx were prevented particularly because of this fencing (Case C-308-08, para. 47).

It is important to emphasize that the above discussion does not exclude effective wilderness protection through existing nature protection conventions and regional systems, such as the EU Natura 2000-regime. If there is a political will to protect wilderness, the existing systems certainly provide excellent opportunities. Wilderness protection may well go hand-in-hand with wetland protection or with the protection of certain species listed in the Bonn or Bern Convention. World heritage sites may also include areas with outstanding wilderness qualities, such as the Tasmanian wilderness. And also the Natura 2000-regime certainly leaves space for wilderness protection (EC Wilderness Guidelines 2013). For instance, in relation to the above mentioned wind turbine case, the Court of Justice of the EU explicitly concluded *"that the Birds and Habitats Directives, in particular Article 6(3) of the Habitats Directive, do not preclude a more stringent national protective measure which imposes an absolute prohibition on the construction of wind turbines [...] within areas forming part of the Natura 2000 network [...]"* (Case C-2/10, para. 58). The weakness discussed above is that generally such protection of wilderness areas is not the main focus of most current international and regional legal regimes and that the protective provisions of these regimes do not specifically require attention for wilderness qualities.

8.4.2 Procedural and Vague Obligations Leave Space for Prioritizing Short-Term Interests

Many provisions and obligations in nature conservation law have a procedural character. Examples include obligations to develop policy plans, to disclose certain activities, to cooperate with other parties, to assess environmental impacts of plans and projects and to monitor change. These obligations have several advantages, but do not contain clear standards of what activities and related influences on natural values are to be considered acceptable. Furthermore, those provisions in nature conservation law that do include more substantial standards are often characterized by vague formulations. This approach makes it possible to reach consensus among a large group of state governments (Birnie et al. 2009, p. 617) and it supports the 'living document'-idea behind many conventions in the sense that the interpretation of the provisions may be adjusted to new challenges and circumstances; however, a major weakness that is directly connected to these advantages is that legal obligations and prohibitions leave so much room for interpretation that in practice, short term economic interests are often prioritized over natural values. In other words, in balancing interests, governments may decide to sacrifice natural values to economic plans and projects without clear violations of the relevant legal instruments. This is in fact also the weak side of the ideal of sustainable development. Balancing interests is in the center of this ideal, however, in practice it also leaves space for prioritization, and often, safeguarding natural values are in a weak position compared to short-term economic interests. Often this results in weak sustainability approaches (Pearce et al. 1989) in which the limitation of adverse impacts on nature ('doing less bad') is considered sufficient for labeling the plan or project as 'sustainable'. This weakness is particularly relevant for wilderness protection because protecting wilderness qualities may often require a more stringent system of prohibiting human activities. For instance, building a hotel in a wetland with high wilderness qualities may under certain conditions (e.g., use of green energy, green waste management, etc.) still be considered within the boundaries of 'wise use', even though the wilderness qualities are likely to be affected.

8.4.3 *Individual Rights to Develop and Non-use of Nature as the Big Taboo*

The previous characteristic is even more problematic because particularly in many Western states the process of modernization and liberalization of the last 200 years has resulted in a great emphasis on the right of the individual to 'develop' and to accumulate wealth through continued appropriation of private property (Linklater 2013). This development is complex and has its roots in many legal, philosophical and economic theories, including those discussed in Sect. 8.2. For instance, the right to appropriate natural components by mixing it with your labour was acknowledge by Locke "as much as any one can make use of to any advantage of life before it spoils" (Locke 1690, para. 31). Consequently, taking more than a person would use, would be spoilage and against the will of God; however, as explained by MacPherson, Locke considered this 'spoil-limitation' for acquisition not relevant anymore after the introduction of money:

> Gold and Silver do not spoil; a man may therefore rightfully accumulate unlimited amounts of it, 'the exceeding of the bounds of his just Property not lying in the largeness of his

> Possession, but the perishing of anything uselessly in it (MacPherson 1962, p. 204; Bell et al. 2004).

This line of reasoning has been strengthened by economic theories, such as the 'invisible hand'-theory of Adam Smith (Smith 1759, p. 203, Van Heerikhuizen 2012). If individuals in a society act in the benefit of their own interests, this will also be best for society: *"By pursuing his own interest he [a person] frequently promotes that of the society more effectually than when he really intends to promote it"* (Smith 1776, p. 562). Such legal and economic theories are further strengthened by social-psychological theories. For instance, Veblen explained at the end of the nineteenth century that the concept of private property results in a competition within society that is based on comparison and imitation: *"The motive that lies at the root of ownership is emulation. [...] The possession of wealth confers honour; it is an invidious distinction"* (Veblen 1899, chapter 2). Over time, private property has increasingly been viewed *"as evidence of the prepotence of the possessor of these goods over other individuals within the community. The invidious comparison now becomes primarily a comparison of the owner with the other members of the group"* (Veblen 1899, chapter 2). Veblen explains that, as a consequence, in Western societies accumulation of wealth even becomes a necessity:

> With the growth of settled industry, therefore, the possession of wealth gains in relative importance and effectiveness as a customary basis of repute and esteem. [...] It therefore becomes the conventional basis of esteem. Its possession in some amount becomes necessary in order to [achieve] any reputable standing in the community. It becomes indispensable to accumulate, to acquire property, in order to retain one's good name (Veblen 1899, chapter 2).

This theory relates very well with recent socio-psychological research that shows that selfish behaviour of individuals in society is not so much motivated by selfishness, but rather by the desire of the individual to prevent a weak position in society (De Dreu 2010, p. 11).

Although this discussion is of course far from complete, such historic views and theories, the strong belief in a liberal market economy and the related views on a limited role of government will at least in part explain why today many governments appear to have difficulties with saying 'no' to plans and projects for reasons of nature conservation. In discussions on the acceptability of human plans and projects there appears to be a general starting point that any person may conduct any activity at any time and place, without firm requirements to prove the importance of the initiative for society. The government often has the burden of proof to demonstrate why a private initiative should not take place and it appears that nature conservation (and probably particularly wilderness protection) is often too weak to overcome that burden. Moreover, also within governments, for instance in processes of developing policy and taking decisions on permit applications, short-term interests are often prioritized. Certainly, anno 2016, the intensifying environmental concerns have made it acceptable that activities are subjected to procedural requirements (e.g. environmental impact assessment) and to certain conditions to limit adverse impacts on the environment, but denying authorization still appears to be a taboo.

This weakness, that generally limits the effectiveness of nature conservation, is particularly problematic for wilderness protection, as wilderness requires not just a balancing of interests, but rather a strict prohibition of human activities that would affect the wilderness qualities.

8.4.4 No Answer to the Question of Accumulation of Adverse Impacts

The above discussed focus on biodiversity and sustainability in international and regional nature conservation law, the substantial space within these systems to balance interests and the government's difficulty to say 'no' to private initiatives, jointly result in a situation in which many human activities are considered to be acceptable or which are explicitly authorized, while they still have a certain adverse impact on nature. Certainly, due to environmental legislation such impacts are subjected to prior assessments and may have been minimized by permit conditions and regulations; however, it often is the accumulation of all these smaller impacts that causes the greatest concerns. Vöneky refers to Francioni, who has stated that *"most environmental damage is caused by lawful acts that have had adverse effects on the environment"* (Vöneky 2008, p. 176–177; Francioni 1994, p. 223) and this problem has become even more apparent over the last two decades. Most of the serious concerns for nature are caused by accumulative impacts of 'lawful' activities, activities that also grow in number, intensity and geographical scope. At the global scale we may refer to climate change and the over-exploitation of certain minerals and at the regional level examples include overexploitation of fish and fresh water stress. These examples may also be relevant at the domestic or even local level, in parallel to many other examples of accumulative problems, such as nitrate deposition or even the scarcity of space. The international conventions and implementing legislation appear to leave much space for allowing individual activities with low levels of adverse effects, while in the end the accumulative impacts are serious hurdles for reaching the conservation objectives.

An example of this weakness, directly relevant for wilderness protection, relates to the establishment of research stations and logistic infrastructure in Antarctica (Bastmeijer 2009). The Protocol on Environmental Protection to the Antarctic Treaty is one of the very few international treaties that provide wilderness values an explicit recognition. Article 3(1) of the Protocol provides an overview of all values that must be taken into account when planning and conducting human activities in the Antarctic and among these values are also the "intrinsic value of Antarctica, including its wilderness and aesthetic values." One of the consequences of this provision is that wilderness values must be taken into account when making an environmental impact assessment in accordance with Article 8 and Annex I to the Protocol. In practice, however, these values are often not receiving serious attention, even by the state governments that are Contracting Party to the Protocol. This may be illustrated by the discussion in the Committee for Environmental Protection (CEP, an advisory body of the Parties to the Treaty and the Protocol) in 2004 on the draft-comprehensive environmental evaluation (Bastmeijer and Roura 2008) for the Czech Republic research station, referred to above:

> New Zealand suggested that, with respect to wilderness values, there are alternatives to building a base on an island where there is no base. […] The Czech Republic advised that they acknowledge the impacts that the base would likely have on wilderness values, but in following the Madrid Protocol they focused on the impact on measurable factors, and contend that on this basis the likely environmental effects of the project are acceptable. They noted that the concept of wilderness values is very philosophical and difficult to quantify objectively, and possibly of greater relevance to the consideration of tourism activities (CEP 2004, para. 53).

Such views result in a practice in which wilderness values receive little weight in balancing interests and in the decision making process. Although wilderness values receive explicit acknowledgement in this legal system, there is clearly sufficient space for balancing interests and, eventually, for considering the project and its adverse impacts "acceptable". As has been shown by Summerson, Tin and others, this has resulted in a substantial accumulation of over 100 research stations (COMNAP 2013) and over 620 'items of infrastructure' in Antarctica, including airstrips, transport facilities and storage facilities (Summerson 2012, p. 89; Tin and Summerson 2009; Carver and Tin 2013).

8.5 Conclusion: The Importance of Deliberate Choices and Drawing Lines on Maps

The strong belief in past centuries in the Western world that the earth is meant for humankind and the legal doctrines that appropriation of nature as private property and territorial land claims of states were only justified in case of actual occupation and use, may be considered as stimuli for the transformation of the earth surface and makes clear that wilderness protection is not embedded in our Western legal roots (Macnab 2009). While this transformation process continued to intensify due to population growth, labour division, technological development, and other factors, the downside of economic development became increasingly clear. Particularly since the nineteenth century attitudes towards nature have been changing and much law has been developed to protect nature; however, there are several weaknesses in global legal systems that substantially limit the role of law in protecting nature, and particularly in protecting wilderness.

We may try to address these weaknesses through small adjustments in the legal system; however, the main message of the previous section is that, most likely, the weaknesses have much more to do with weaknesses of humankind itself, such as the difficulty to accept limitations to our social and economic ambitions and our disability to deal with accumulative impacts. These problems are also clearly reflected in quite some other global environmental problems, such as marine litter, biodiversity loss and climate change. If we think there should be wild places left in this world for present and future generations of humankind and other species, it is important to make the explicit decision to ensure this. As concluded by Dearden, *"[i]f wilderness remains on this planet one hundred years from now it will be because, for the first time in the history of man, we have deliberately chosen that it should be so as a positive benefit rather than an industrial remnantn* (Dearden 1989, p. 206). Without such an explicit decision, the above-described weaknesses make a further decline of wilderness most likely.

For making such deliberate choices, it is important to map existing wilderness and to 'draw lines on maps': identify and designate areas for wilderness protection. To implement such deliberate choices, it is also important to have a clear picture of the overlap between such areas with high wilderness qualities and areas that are already protected under legal systems. For those wilderness areas that already have a protected status, it is important to understand to what extent this protection is also aimed at protecting wilderness qualities. This is important as wilderness qualities are often not among the criteria for selection, designation and protection under existing conventions. Along these lines it may be necessary to broaden existing legal protection to ensure wilderness protection, as well as designating wilderness areas that currently do not have any legally protected status.

To ensure long term effects at the regional and global scale, this approach could best be worked out in clear, legally binding agreements between states. This could be done within the frameworks of existing international nature conservation conventions. Even though wilderness protection is not an explicit policy objective under most of these conventions, protecting wildernesses may go hand-in-hand or may even be crucial for achieving other targets. As it is uncertain whether the international community is capable of making such deliberate choices, state governments may also make such policy choices at the domestic level. Such initiatives may be built on best practices in other countries. For instance, in 2014 the 50th anniversary of the U.S. Wilderness Act 1964 will be celebrated and there are many other inspiring examples in other regions (Kormos and Locke 2008). Under these domestic regimes various wild, relatively undisturbed areas within the territory of these states have been designated and protected by law to ensure that these areas do not lose their wilderness characteristics due to human activities.

Would this result in separating humans from nature, a criticism that has often been expressed in the past in relation to wilderness protection (Cronon 1995)? Not necessarily. Drawing lines on maps is something different than fencing geographical areas to keep people out. Human activities that do not cause adverse impacts on the wilderness qualities of the area may well be allowed. Enjoyment of these wilderness qualities by people is often even an explicit aim of existing domestic wilderness legislation. The real separation of humans and nature takes place due to the fast global process of urbanisation, and access to nature in and around cities, as well as in wild places, is important for our understanding and appreciation of nature.

Acknowledgement The author would like to thank Tim van der Putten for his proofreading and editorial support.

References

Aristotle. (350 b.c). *The politics, A treatise on government*. translation from the Greek original by William Ellis, London & Toronto: J.M. Dent & Sons Ltd./New York: E.P. Duton &. Co, first issue in 1912, reprinted in 1919, 1923, 1928. Available at: www.gutenberg.org/files/6762/6763-h/6762-h.htm

Baakman, K. (2011). *Testing times: The effectiveness of five international biodiversity-related conventions*. Nijmegen: Wolf Legal Publishers.

Balch, T. W. (1910). The arctic and Antarctic regions and the law of nations. *American Journal of International Law, 4*, 265–275.

Bastmeijer, C. J. (2009). Protecting polar wilderness: Just a western philosophical idea or a useful concept for regulating human activities in the polar regions? In G. Alfredsson, T. Koivurova, & D. Leary (Eds.), *The yearbook of polar law* (Vol. 1, pp. 73–99). Boston/Leiden: Brill/Martinus Nijhoff Publishers. Available at: http://ssrn.com/abstract=1295430

Bastmeijer. C. J., & Roura, R. (2008). Environmental impact assessment in Antarctica. In C. J. Bastmeijer, & T. Koivurova (Eds.), *Theory and practice of transboundary environmental impact assessment* (pp. 175–219). Leiden: Brill/Martinus Nijhoff Publishers. Available at: http://ssrn.com/abstract=1297285

Bastmeijer, K. (2016). Introduction: An international history of wilderness protection and central aim of this book. In K. Bastmeijer (Ed.), *Wilderness protection in Europe. The role of international, European and national law*. Cambridge: Cambridge University Press.

Bell, S. A., Henry, J. F., & Randall Wray, L. (2004). A chartalist critique of John Locke's theory of property, accumulation, and money: Or, is it moral to trade your nuts for gold? *Review of Social Economy, 62*(1), 51–65. Available at: http://www.tandfonline.com/doi/pdf/10.1080/0034676042000183826

Birnie, P., Boyle, A., & Redgewell, C. (2009). *International law & the environment*. Oxford: Oxford University Press.

Bowland, M. J. (1988–1989). The protection of animals under international law. *Connecticut Journal of International Law, 4*, 487–499.

Caddell, R. (2005). International law and the protection of migratory wildlife: An appraisal of twenty-five years of the Bonn convention. *Colorado Journal of International Environmental Law & Policy, 16*(113), 113–156. Available at: http://heinonline.org/HOL/LandingPage?handle=hein.journals/colenvlp16&div=10&id=&page

Carver, S. (2016). Mapping wilderness in Europe. In K. Bastmeijer (Ed.), *Wilderness protection in Europe: The role of international, European and national law*. Cambridge: Cambridge University Press.

Carver, S., & Tin, T. (2013). *Mapping and modelling wilderness values in Antarctica*. Leeds: Wildland Research Institute, University of Leeds. Available at http://www.ats.aq/documents/ATCM36/att/ATCM36_att116_e.pdf

CEP. (2004). *Report of the VIIth meeting of the CEP, Cape Town*. Available at: http://cep.ats.aq/cep/MediaItems/atcm27_cepvii_e.pdf

Cliteur, P. (2001). *Darwin, Dier en Recht*. Essay available at: http://media.leidenuniv.nl/legacy/Darwin%20%20dier%20en%20recht.pdf

Cliteur, P. (2005). De filosofie van dierenrechten. In J. Braeckman, B. de Reuver, & T. Vervisch (Eds.), *Ethiek van DNA tot 9/11* (pp. 135–159). Amsterdam: Amsterdam University Press. Available at: http://media.leidenuniv.nl/legacy/De%20filosofie%20van%20de%20dierenrechten.pdf

COMNAP. (2013). *Main Antarctic facilities operated by the National Antarctic programs in the Antarctic treaty area*. Available at: https://comnap.aq/Members/Shared%20Documents/Antarctic_Facilities_list_15May2013.xls

Cronon, W. (1995). The trouble with wilderness; or, getting back to the wrong nature. In W. Cronon (Ed.), *Uncommon ground: Rethinking the human place in nature*. New York: W.W. Norton &

Co. Available at http://www.warren-wilson.edu/~ghawkins/radical/docs/Cronon_Trouble_w_Wilderness.pdf

Darwin, C. (1871). *The descent of man, and selection in relation to sex*. London: John Murray.

De Dreu, C. (2010). Verbinden als sociaal dilemma. *Idee, 31*, 10–13.

De Vattel, E. (1758). *Le Droit des gens* (Quotations taken from the English translation by Joseph Chitty (1867): *The Law of Nations*). Philadelphia: Johnson & Co.

Dearden, P. (1989). Wilderness and our common future. *Natural Resources Journal, 29*, 205–221.

Deinet, S., Ieronymidou, C., McRae, L., Burfield, I. J., Foppen, R. P., Collen, B., & Böhm, M. (2013). *Wildlife comeback in Europe: The recovery of selected mammal and bird species. Final report to Rewilding Europe by ZSL, BirdLife International and the European Bird Census Council.* London: ZSL. Available at: http://static.zsl.org/files/wildlife-comeback-in-europe-the-recovery-of-selected-mammal-and-bird-species-2576.pdf

DeMello, M. (2012). *Animals and society. An introduction to human-nature studies*. New York: Columbia University Press.

Doremus, H. (1999). The special importance of ordinary places. *Environmental Law & Policy Journal, 23*(3). Available at http://scholarship.law.berkeley.edu/facpubs/511

Dudley, N. (Ed.). (2008). *Guidelines for applying protected area management categories*. Gland: IUCN. Available at: http://data.iucn.org/dbtw-wpd/edocs/2008-028.pdf

Fisher, M. (2016). Ecological values of wilderness in Europe. In K. Bastmeijer (Ed.), *Wilderness protection in Europe: The role of international, European and national law*. Cambridge: Cambridge University Press.

Francioni, F. (1994). Liability for damage to the common environment: The case of Antarctica. *RECIEL, 3*, 223.

Goodricht, G. (1951). The United Nations conference on resources. *International Organization, 5*(1), 48–60.

Gorter, H. P. (1956). Fifty years 'Natuurmonumenten' (nature reserves). In *Vijftig jaar natuurbescherming in Nederland* (pp. 11–66). Amsterdam: Drukkerij De Volharding. Commemorative book on the occasion of the fiftieth anniversary of the Society for the Protection of Nature (in the Netherlands).

Grotius, H. (1609). *Mare Liberum, The Freedom of the Seas* (Translated with a revision of the Latin text of 1633 by Van Deman Magoffin, R. (1916)). New York: Oxford University Press.

Huxley, T. H. (1863). *Evidence as to man's place in nature*. New York: D. Appleton and Company. Also available at https://archive.org/details/evidenceastomans00huxl

IUCN. (2004). *International environmental governance: An international regime for protected areas*. Gland: IUCN Environmental Policy and Law Paper 49.

Jeffery, M. I. (2004). An international legal regime for protected areas. In John Scanlon, & Burhenn-Guilmin (Ed.), *International environmental governance* (pp. 9–40). Gland: IUCN. Available at https://portals.iucn.org/library/efiles/documents/EPLP-049.pdf

King Hubbert, M. (1969). Energy Resources. In National Research Council – Division of Earth Sciences – Committee of Resources and Man. *Resources and man* (Vol. 238). San Francisco: W.H. Freeman and Company.

Koester, V. (2012). Book review of Baakman's 'testing times: The effectiveness of five international biodiversity-related conventions'. *RECIEL, 21*(1), 67–70.

Kormos, C. F., & Locke, H. (2008). Introduction. In C. F. Kormos (Ed.), *A handbook on international wilderness law and policy*. Colorado: Fulcrum Publishing, Golden.

Kuiters, A. T. et al. (2012). *Guidelines for the management of wilderness and wild areas in Natura 2000*. Brussels: European Commission. Available at: http://portal.icnb.pt/NR/rdonlyres/773250A6-7A20-4F4E-A944-255E90D82E73/0/GuidelinesWilderness_2012.pdf

Linklater, A. (2013). *Owning the earth: The transforming history of land ownership*. New York: Bloomsbury.

Locke, J. (1690). *The second treatise of government*. Available at: http://libertyonline.hypermall.com/Locke/second/second-frame.html

Lowenthal, D. (2000). Nature and morality from George Perkins Marsh to the millennium. *Journal of Historical Geography, 26*(1), 3–27. Available at: http://geography.fullerton.edu/taylor/ENST595T/Lowenthal_JHG.pdf

Macnab, R. (2009). "Use it or lose it" in Arctic Canada: Action agenda or election hype? *Vermont Law Review, 34*, 2–14. Available at http://lawreview.vermontlaw.edu/files/2012/02/macnab.pdf

MacPherson, C. B. (1962). *The political theory of possessive individualism – Hobbes to Locke*. Oxford: Oxford University Press.

Marsh, G. P. (1867). Preface of *man and nature*. New York: Charles Scribner & Co. Available at http://www.gutenberg.org/files/37957/37957-h/37957-h.htm

Minteer B. A., Manning R. E. (2005). An appraisal of the critique of anthropocentrism and three lesser known themes in Lynn White's "The historical roots of our ecological crisis". *Organization & Environment 18*(2005), 163–176 Also available at: https://www.uvm.edu/parkstudieslaboratory/publications/Appraisal_Critique_Anthropocentrism.pdf

Morgera, E., & Tsioumani, E. (2011). *Yesterday, today and tomorrow: Looking afresh at the convention on biological diversity* (University of Edinburgh School of Law Working Paper No. 2011/21). Available at: http://ssrn.com/abstract=1914378

Nash, R. F. (2001). *Wilderness & the American mind* (4th ed.). New Haven: Yale University Press.

Passmore, J. (1974/1980). *Man's responsibility for nature*. London: Duckworth.

Pearce, D. W., Markandya, A., & Barbier, E. (Eds.). (1989). *Blueprint for a green economy* (Vol. 1). London: Earthscan.

Roberts, N. (1996). The human transformation of the earth's surface. *International Social Science Journal, 150*, 493–510.

Sands, P. (1995). *Principles of international environmental law*. Manchester: Manchester University Press.

Schrijver, N. (2008). *The evolution of sustainable development in international law: Inception, meaning and status, pocketbook of the Hague Academy of International Law*. Leiden/Bosten: Martinus Nijhoff Publishers.

Schrijver, N., & Prislan, V. (2009). From *Mare Liberum* to the global commons: Building on the Grotian heritage. *Grotiana, 30*, 168–206.

Secretariat of the Convention on Biological Diversity. (2010). *Global biodiversity outlook 3*. Montréal. Available at: http://www.cbd.int/doc/publications/gbo/gbo3-final-en.pdf

Smith, A. (1759). *Glasgow edition of the works and correspondence, Vol. 1 The theory of moral sentiments*. In D. D. Raphael, & A. L. Macfie (Eds.), (1982). Indianapolis: Liberty Fund. Available at: http://files.libertyfund.org/files/192/Smith_0141-01_EBk_v6.0.pdf

Smith, A. (1776). *An inquiry into the nature and causes of the wealth of nations* (S. M. Soares, Ed.). (2007). MetaLibri Digital Library.

Snyder, D. C. (2007). Locke on natural law and property rights. In T. Brooks (Ed.), *Locke and law, philosophers and law series* (pp. 3–30). Hampshire: Ashgate.

Summerson, R. (2012). Protection of wilderness and aesthetic values in Antarctica. In F. Huettmann (Ed.), *Protection of the three poles* (pp. 77–109). Dordrecht: Springer.

Thomas, K. (1983). *Man and the natural world, changing attitudes in England (1500–1800)*. Harmondsworth/Middlesex: Allen Lane/Penguin Books Ltd.

Tickell, C. (1993). The human species: A suicidal success? *The Geographical Journal, 159*(2), 219–226. Available at: http://www.jstor.org/stable/3451413

Tin, T., & Summerson, R. (2009). *Wilderniss and aesthetic values in Antartica*. Map presented by ASOC at the ATCM XXXII.

Trouwborst, A. (2008). Seabird Bycatch – Deathbed conservation or a precautionary and holistic approach? *Journal of International Wildlife Law & Policy, 11*(4), 293–333.

Van Heerikhuizen, B., (2012). *An academic lecture on Adam Smith and Bernard de Mandeville*, Available at http://www.youtube.com/watch?v=A8wMy1JvSsY

Van Heijnsbergen, P. (1997). *International legal protection of flora and fauna*. Amsterdam: Ios Press.

Veblen, T. (1899). *The theory of the leisure class. An economic study of institutions*. New York: Macmillan Company. (1909 ed. freely available at http://www.gutenberg.org/files/833/833-h/833-h.htm)

Verbeek, T. H. M. (2007). *Mechanistische filosofie: Descartes en Spinoza*. [Online] Available at: http://igitur-archive.library.uu.nl/sg/2007-1115-200507/3-mechanis.pdf

Vöneky, S. (2008). The liability annex to the protocol on environmental protection to the Antarctic treaty. In D. König, P. T. Stoll,, V. Röben, & N. Matz-Lück (Eds.), *International law today: New challenges and the need for reform? Beiträge zum ausländischen öffentlichen Recht und Völkerrecht*. Berlin/Heidelberg/New York: Springer. Available at: http://www.mpil.de/shared/data/pdf/text_liabilityannex_voeneky.pdf

Wells, S. (2010). *Pandora's seed: The unforeseen cost of civilization*. New York: Random House Publishers. Translated into Dutch as *De akkers van Pandora*. Amsterdam: Uitgeverij Contact 2010.

White Jr., L. (1967). The historical roots of our ecological crisis. *Science, 155*(3767). 1203–1207. Available at http://www.drexel.edu/~/media/Files/greatworks/pdf_fall09/HistoricalRoots_of_EcologicalCrisis.ashx

Wild Europe Initiative. (2013). *A working definition of European wilderness and wild areas*. Availableat:http://www.wildeurope.org/images/pdf/a-working-definition-of-european-wilderness-and-wild-areas.pdf

World Resources Institute. (2005). *Millennium ecosystem assessment*. Washington, DC: Synthesis. Available at: www.maweb.org/documents/document.297.aspx.pdf

Zweers, W. (1995). *Participeren aan de natuur; ontwerp voor een ecologisering van het wereldbeeld*. Amsterdam: Jan van Arkel.

Law References

Bird Directive. (1979). Directive 79/409/EEC on the conservation of wild birds, OJ L 103/1.

Convention for the Protection of Wild Animals, Birds and Fish in Africa. (1900). Available at: http://iea.uoregon.edu/pages/view_treaty.php?t=1900-PreservationWildAnimalsBirdsFishAfrica.EN.txt&par=view_treaty_html

Convention on Biological Diversity. (CBD), Rio de Janeiro, 1992. Available at: http://www.cbd.int/convention/text/

Convention on the conservation of European wildlife and natural habitats Bern. (1979). Available at: http://conventions.coe.int/Treaty/en/Html/104.htm

Convention on the conservation of migratory species of wild animals. (1979). *Bonn*. Available at: http://www.cms.int/documents/convtxt/cms_convtxt_english.pdf

Convention on Wetlands of International Importance. (1971). Ramsar. Available through: www.ramsar.org

Convention to Protect Birds Useful to Agriculture. (1902). *Parry's consolidated treaty series, 191*(91). Available at: http://eelink.net/~asilwildlife/bird_1902.html

European Commission. (2013). *Guidelines on wilderness in Natura 2000. Management of terrestrial wilderness and wild areas within the Natura 2000 network* (Technical report 2013 – 069). Brussels: European Union. Available at: http://ec.europa.eu/environment/nature/natura2000/wilderness/pdf/WildernessGuidelines.pdf

European Court of Justice of the EU, Case C-127/02 *Landelijke Vereniging tot Behoud van de Waddenzee and Nederlandse Vereniging tot Bescherming van Vogels v Netherlands* [2004].

European Court of Justice of the EU, Case C-308-08 *European Commission v Kingdom of Spain* [2010].

European Court of Justice of the EU, Case C-2/10 *Azienda Agro-Zootecnica Franchini Sarl and Eolica di Altamura Srl v Italy* [2011].

Habitat Directive. (1992). Directive 92/43 on the Conservation of natural habitats and wild fauna and flora, OJ L 206/7.

Protocol on Environmental Protection to the Antarctic Treaty. (1991). *ILM* 30(1455). Also available at http://www.ats.aq/e/ep.htm

Stockholm Declaration, adopted at the UN-Stockholm Conference on the Human Environment in 1972, available through http://www.unep.org/

Chapter 9
Unraveling the Coil of the Wild: Geospatial Technology and Wilderness

Mark L. Douglas and William T. Borrie

Abstract The question of wilderness and technology has been widely debated in the human dimensions of natural resources, with scholars showing the impacts of technology upon the wilderness experience. However, these works have not fully addressed the impact of geographic information system technology on wild space. Our chapter addresses the impact of GIS on wild space with special attention to how wild space is misunderstood as mere resource to be discretely ordered and regulated. Specifically, this chapter looks at both the existential thinking of Martin Heidegger and more contemporary social-ecological work. We show how geospatial technology surreptitiously divides information about wild space from being in wild space. We discuss the phenomena of Google Trekker and remote viewing, and juxtapose them against being in wild space. We dismiss the misconception that geospatial representation does justice to the focal space held by wilderness. We argue that specific danger lies in any blind assumption that geospatial wilderness data exhausts the truth of wild space. In conclusion, this chapter, by closely examining the systemization of wild space information, sheds new light on the potentially overlooked difference between understanding wildness and understanding geospatial wilderness attributes.

Keywords Non-representational space • Wildness • Google Trekker • Wilderness experience

9.1 Introduction

Technology generally changes both the meaning and experience of wilderness (Borrie 1998; Freimund and Borrie 1998; Martin and Pope 2012; Shultis 2012). By design, technology makes life easier, more convenient and more controlled.

M.L. Douglas (✉)
Environmental Recreation and Tourism Management Program,
University of Maine at Machias, Machias, ME, USA
e-mail: mark.douglas@maine.edu

W.T. Borrie
College of Forestry and Conservation, University of Montana, Missoula, MT, USA

S.J. Carver, S. Fritz (eds.), *Mapping Wilderness*,
DOI 10.1007/978-94-017-7399-7_9

As Pohl (2006) points out, technological devices may detract from the very purposes for which people visit wilderness. When technology is imposed on wilderness, either heedlessly or for external purposes, the range of practices and experiences can be reduced. Those technologies, "help us control our surroundings, shift our focus to consumption and materialism, dominate nature, and remove us from bodily engagement" (p. 159). Wilderness is the counterpoint to technological culture.

How does geospatial technology make an impact? The original structure, significance, and mystery of wilderness are transformed when wild landscapes are digitally recorded, rendered and stockpiled on data servers. Data grids and ubiquitous virtual access prescribe the relationship between humans and the wild world. For example, with projections that rely on geodesy, spatial technologies convert the earth to a flat surface. This mirrors the concerns expressed by philosopher Martin Heidegger who noted that this, *"projection sketches out in advance the manner in which the knowing procedure must bind itself and adhere to the sphere opened out"* (1977a, p. 118). In this fashion, human knowledge and understanding of the world is bound to the structures of the technology. Spatial engagement with the qualities of wilderness, in terms of science, is mediated by the technological platform.

Geospatial technology is a translation of the world into representational figures, and again as Heidegger foreshadowed, *"where the world becomes picture, what is, in its entirety, is juxtaposed as that for which man is prepared and which, correspondingly, he therefore intends to bring before himself and have before himself, and consequently intends in a decisive sense to set in place before himself"* (p. 129). The technology is, therefore, an un-wilding of the world, a bringing of it into, and a positioning under the discretion of human purpose. Humanity imposes its conception of the world and imposes an anthropocentric will. The primeval character and influence of wild volition get reduced through human and mechanical conscription. In contrast, the character of wild places is unknown-ness or mystery. Our experiences of these places have been described as encompassing "unintelligibility" (Malpas 2007). Wilderness means more than what can be represented in the visual spectrum, mathematically, or on a screen.

It is our position that geospatial technology increases the quantity of wilderness information while risking the quality of wilderness understanding. Safeguarding wilderness character calls less for supremely accurate surveys and more for indiscreet wonder, *"a qualification of disinterestedness with which human interest requires to be informed"* (Bugbee 1974, p. 616). We may withdraw into the simple presence of wildness, held by wild events, instants, and moments. *"Wildness is uncaged, self-willed, self-governing, and not subject to the impositions of another"* (Borrie et al. 2012, p. 71). Wildness is uncanny. We simultaneously welcome geospatial science and honor the mystery of wilderness. This chapter examines the difference between building stores of wilderness data and deepening a sense of the wild.

As a specific example, to which we will return throughout, we consider the Google Trekker program launched recently at a media event in Grand Canyon National Park. The device uses an array of fifteen lenses along with, "accelerometers, gyros, [a] magnetometer as well as GPS" to visually appropriate the surrounding environment along hiking trails (Henn 2012). While we do not suggest that Google Inc. intends an experiential equivalence between sauntering along the trail in the flesh and virtual hiking via internet access, they do tout delivering "the opportunity to marvel at this beautiful, majestic site from the comfort of their computers or mobile devices" (Falor 2012). How this technology changes what is knowable, and how it exchanges a meaningful context for data, are important considerations. How does geospatial technology like Google Trekker render a wild landscape as vast and magnificent as the Grand Canyon? The works of Martin Heidegger, and those who have studied his writing, offer an avenue and standpoint from which to consider this question.

9.2 Heidegger's Investigation of Technology

Heidegger (1977b) has eight points to understand how technology renders things.

(1) Technology discloses things in a particularly modern-era way. For instance, the Cartesian grid of latitude and longitude was designed for particularly instrumental use and positivist scientific thinking. The Cartesian grid puts the Earth into a particular form of representation. Technology highlights *what* a human wants to see (an object) and *how* humanity seeks to know the world (objectively), and thus only those ideas and claims that match the purposes and approaches of positivistic science get illustrated in our maps.

(2) Technology challenges and provokes when it engages and renders things. Devices bring about a setting upon or seizing of nature. In bringing the wild perplexity of nature into the scope of human knowledge systems, complexities must be corralled and tamed to allow for human management. This is simplification and reduction. The Grand Canyon, for instance, with its labyrinth of side canyons and vast relief perplexes some who encounter it despite their adoption of the latest device. Geospatial technologies reduce and simplify the ancient and wrinkled landscape and so, the power and furtive fortitude of the wild world is ruptured.

(3) Technology renders things as an expediting, ordering and stockpiling. Google Trekker will photograph, illustrate and make accessible the mysteries and secrets of the Grand Canyon paths. What was previously veiled, or what demanded engaging conviction, will become merely seen and evident. Clear, concise, and comprehensive efforts continue toward disclosure of whatever veiled wild space abides. Wilderness gets ready-made to order in the on-demand fashion of the times. The changing of meaning opened up with this seamless, simple, and instant availability, we fret, may be ignored or forgotten by Google or resource managers.

(4) Technology renders things as objects, positioned as if "on call for duty" and "on demand" (pp. 17, 18). Wilderness is reduced to its qualities that are compiled in attribute tables and racked up to be called upon "at your service". This occurs

without regard for the many and different meanings that exist among the wild beings. Any limitations of the technology, such as no capacity to translate the circumstantial play of light, sound, texture, odor, and atmosphere, will incapacitate the characterization of the place. Given that geospatial technologies may assume the dominant or only record of place, much gets left out that is essential and possibly significant. Heidegger describes technological transcription as wrenching away the moments when *"man opens his eyes and ears, unlocks his heart, and gives himself over to meditating and striving, shaping and working, entreating and thanking"* thereby replacing those stellar events with instants *"when man, investigating, observing, ensnares nature as an area of his own conceiving"* (p. 19). The richness and the heart of a place are lost in translation. Transcription also discounts moments when other beings interrelate beyond the human discretion (Skocz 2005).

(5) Technology plots things in definitive location, capturing objects in a particular time and place. The recording of the movement (or vibration) of things across time is less common. For example, the geospatial model of the Grand Canyon in 2012 will differ from a future model, but neither deep geologic time nor current flux is likely to be taken into account by a geospatial perspective. The models generated will have a permanency and crystallization that may be neither warranted nor sufficiently true.

(6) Technology is a requisitioning and categorizing force. With the prevalent conception of technology as a neutral force, people assume that the downside of technology depends on how it is applied. This claims that the force of technology depends on how it is wielded as a tool. If technology is categorized as merely a tool, then the categorizing and ordering force of technology is fully in play. Like Snyder (1990), we note the distinct connection between wildness and freedom and feel that categorizing wilderness is a threat to freedom since the freedom to be categorized or not is itself a fundamental freedom. It may seem inevitable that as humans we categorize whatever presents itself. But, avoiding being labeled and freely being away from the control of categories constitutes being in wilderness. *"Freedom is that which conceals in a way that opens to light, in whose clearing there shimmers that veil that covers what comes to presence of all truth and lets the veil appear as what veils"* (Heidegger 1977b, p. 25). In other words, in the freedom of wilderness, a possibility space opens up and grants clearance to freedom from categorization.

(7) The ordering force of technology surreptitiously erodes the human responsibility to guard disposition. Humans have the opportunity to *"be the one who is needed and used for the safekeeping of the coming to presence of truth"* (p. 33). We are unique among the beings of the Earth, with choice and responsibility to approach the truest understanding of the world. Humans have the ability to interpretively model the world and the choice to model and to understand the role of modeling is unique. We humans can also reflexively consider our role in the model. We have a responsibility to consider and understand technological wilderness surveillance and modeling.

(8) Technology brings things under a regulated and secured system. Surveillance induces a stockpiling of resources and is tied to supervision, purpose, or oversight. The geospatial grid, for example, allows for boundaries and artificial demarcations to be imposed, potentially binding the land within regulatory precincts. In similar fashion, there is danger that humans themselves may be seen as mere resources, to be recorded, ordered, and stockpiled. That categorization is a threat to freedom and wildness. In contrast to an increasingly settled, developed, and ordered civilization, wild places provide an escape or sanctuary from impositions and strictures.

9.3 Wilderness and Technology

As a specific application of Heidegger's thinking, David Strong (1995) describes the effects of technology on wilderness. Strong uses Albert Borgmann's device paradigm to show that GIS, as a device, separates the means from the ends of wilderness. That is, the means are the machinery, such as Trekker, that make wilderness safer, easier, more convenient, instantaneous and more ubiquitously available. Wild places become increasingly mundane from only a few keyboard strokes and mouse-clicks. Whereas direct engagement in wilderness shows the place in all its fullness, GIS reduces the setting to something readily called up and then ignored when it is no longer of use. Little effort and responsibility is required, and the consequences of access are few. Every place becomes more like any other place. A uniquely situated sense of a wild place is impoverished and so we lose the opportunity to experience and know its inherent and iridescent power in context.

Jeff Malpas (2007) expands on this idea, suggesting that a full sense of place is neglected as a result of the rise of technology. According to Malpas (p. 16), Heidegger took place to be *"the idea and image of a concrete gathering of otherwise multiple elements in a single unity—as places themselves gathered into a single locality."* That is, place is much more than a mere spatial location or positionality; a wild place is a convergence of many things that are integral to its possibility and actuality (2007). Wilderness is a focal space where many things gather into a whole. Spatial technologies are unlikely to capture the convergence, the focus, or all other contextual contingencies. Things get removed from their context, separated from their associations. It is hard to imagine the full character of wilderness being preserved geodetically.

Further, as Borgmann (1984) writes, technology removes things from the very practice of engaging with those things. Humans have inherited ancestral attunement to natural information, as *"nothing so engages the fullness of human capabilities as a coherent and focused world of natural information"* (Borgmann 1999, p. 219). To access information about the outdoors through other means is both a diminishment of our human experience, but also of the world. With technological proficiency humans are at risk of considering themselves as all-seeing and all-knowing. The wild world is fading into the background with only the data remaining to relate us to the world. Technology transitions us from having information *about* reality to having abstract information *as* reality. As Jean Baudrillard (1994) wrote, the image supplants the original and we lose our ability to distinguish the two. It becomes harder and harder to discern what wilderness preserves, and why we preserve wilderness when we relate to wilderness data instead of that upon which wilderness preservation is premised, wildness. Wilderness holds space for an understanding of wildness more so than an understanding of the space held within wilderness.

In his essay, "Wilderness Ontology," Levi Bryant (2011) further contrasts wilderness against the cosmopolitan city where, *"everything seems to be posited before my knowing or comprehending gaze and everything seems to be arranged for the sake of my instrumental gaze"* (p. 20–21). Is such prepositional comprehension appropriate for wilderness? In contrast to a world of commodities and resources, Bryant encourages wilderness, *"modes of thinking that help us to become attentive to the alterity of things, the thingliness of things"* (p. 23). That is, rather than experiencing a constructed version of wilderness, Bryant encourages engaging wild things as they already are. Rather than a distanced and mediated engagement, he calls for humans to directly engage in wilderness.

This is much like Sabine Hofmeister (2009) who calls for wilderness to change social relations with wild nature. Wilderness is a place that offers *"an opportunity to experience freedom from or lack of orientation and the insecurity that may accompany it"* (p. 310). Wilderness welcomes indiscrete wonder. Wilderness can be a spatially and temporally unspecified, yet immediate, place; a non-normative place that is, *"a type of space beyond socially fixed, functional structures of space and time—an experience of space that may serve to help us to retain our memory of first wilderness"* (p. 310). Wild space is *"the other of society, as something 'outside'"* (p. 302). Wild space is boisterous roistering.

We can see that many of those who have examined the effects of technology on wilderness reinforce the more general concerns expressed by Heidegger. Wilderness is defined as a place in contrast to an increasingly crowded and controlled society. It is a place of freedom, undeveloped, unconstrained, and private. In this paper, we have laid out the distinct threats of geospatial technologies to wilderness. By design, those technologies reduce wildness. Mystery and challenge are reduced and the full vibrancy and richness of wilderness is lost. Geospatial technology, in its distant, all-seeing, and static stance, stands in contrast to an intimate, earned, and responsive engagement with wilderness. The specific danger is that geospatial renderings may be blindly assumed to stand for, or represent, real wilderness.

The application of geospatial science is increasingly called for by decision makers. It offers many advantages for stewardship planning and administration. GIS represents the space of things and herein lays the crux of the matter. Geospatial information is only (and powerfully) a representation and cannot replace or understand real wild space.

9.4 Conclusion

Wild space grants clearance, and as Heidegger says, *"only this clearing grants and guarantees to us humans a passage to those beings that we ourselves are not, and access to the being that we ourselves are"* (2001, p. 51). Only by standing with and within, instead of in opposition to, or by imposition of wild space can we fully appreciate what abounds. To endorse the surveillance way of technology is to deprive ourselves of clearance to the middle ground of being other. Endorsing technological wild space surveillance imposes ourselves ahead of wilderness and deprives us of clearance to the ground of being. We offer a token of caution and care. Our reflexive concern is over heedless imposition of geospatial technology upon wild space. If GIS and devices like Google Trekker squelch human engagement with, and attunement to, wilderness (as we have argued), then the appropriate imposition of geospatial technology upon wild space must be unremittingly reflected upon and mindfully interrogated.

References

Baudrillard, J. (1994). *Simulacra and simulation*. Ann Arbor: University of Michigan Press.

Borgmann, A. (1984). *Technology and the character of contemporary life: A philosophical inquiry*. Chicago: The University of Chicago Press.

Borgmann, A. (1999). *Holding on to reality: The nature of information at the turn of the millennium*. Chicago: The University of Chicago Press.

Borrie, W. T. (1998). Impacts of technology on the meaning of wilderness. In A. E. Watson, G. H Aplet, & J. C. Hendee (Comps.), *Personal, societal, and ecological values of wilderness: Sixth world wilderness congress proceedings on research, management, and allocations*. Proceedings of the RMRS-P-4 (Vol. 2, pp. 87–88). Ogden: US Department of Agriculture, Forest Service, Rocky Mountain Research Station.

Borrie, W. T., Meyer, A. M., & Foster, I. M. (2012, April). Wilderness experience as sanctuary and refuge from society. In D. N. Cole, (Comp.), *Wilderness visitor experiences: Progress in research and management*; 2011 April 4–7; Missoula, MT. Proceedings of the RMRS-P-66 (pp. 70–76). Fort Collins: US Department of Agriculture, Forest Service, Rocky Mountain Research Station.

Bryant, L. R. (2011). Wilderness ontology. In C. Jeffery (Ed.), *Preternatural* (pp. 19–26). Brooklyn: Punctum Books.

Bugbee, H. G. (1974). Wilderness in America. *Journal of the American Academy of Religion, 42*(4), 614–620.

Falor, R. (2012, October 24). *Trekking the Grand Canyon for Google maps*. Retrieved from Google, official blog: http://googleblog.blogspot.com/2012/10/trekking-grand-canyon-for-google-maps.html

Freimund, W. A., & Borrie, W. T. (1998). Wilderness in the 21st century – Are there technical solutions to our technical problems? *International Journal of Wilderness, 3*(4), 21–23.

Heidegger, M. (1977a). The age of the world picture. In *The question concerning technology and other essays* (W. Lovitt, Trans., pp. 115–154). New York: Harper & Row.

Heidegger, M. (1977b). The question concerning technology. In *The question concerning technology and other essays* (W. Lovitt, Trans., pp. 3–35). New York: Harper & Row.

Heidegger, M. (2001). The origin of the work of art. In *Poetry, language, thought* (A. Hofstadter, Trans., pp. 17–86). New York: Harper & Row.

Henn, S. (2012, October 24). *Google's street view goes into the wild*. Retrieved from National Public Radio website: http://m.npr.org/news/Technology/163499448

Hofmeister, S. (2009). Nature's running wild: A social-ecological perspective on wilderness. *Nature and Culture, 4*(3), 293–315.

Malpas, J. (2007). *Heidegger's topology: Being, place, world*. Cambridge: MIT Press.

Martin, S. R., & Pope, K. (2012, April). The influence of hand-held information and communication technology on visitor perceptions of risk and risk-related behavior. In D. N. Cole, (Comp.), *Wilderness visitor experiences: Progress in research and management*; 2011 April 4–7; Missoula MT. Proceedings of the RMRS-P-66 (pp. 119–126). Fort Collins: U.S. Department of Agriculture, Forest Service, Rocky Mountain Research Station.

Pohl, S. (2006). Technology and the wilderness experience. *Environmental Ethics, 28*(2), 147–163.

Shultis, J. (2012, April). The impact of technology on the wilderness experience: A review of common themes and approaches in three bodies of literature. In D. N. Cole, (Comp.), *Wilderness visitor experiences: Progress in research and management*; 2011 April 4–7; Missoula MT. Proceedings of the RMRS-P-66 (pp. 110–118). Fort Collins: U.S. Department of Agriculture, Forest Service, Rocky Mountain Research Station.

Skocz, D. (2005). Wilderness management and geospatial technology: A view from the Black Forest. *Environmental Philosophy, 2*(2), 53–60.

Snyder, G. (1990). The etiquette of freedom. In *The practice of the wild* (pp. 3–26). Berkeley: Counterpoint.

Strong, D. (1995). *Crazy mountains: Learning from wilderness to weigh technology*. Albany: State University of New York Press.

Chapter 10
Wilderness Areas in Romania: A Case Study on the South Western Carpathians

Dragoş Ştefan Măntoiu, Marius Costin Nistorescu, Ionuţ Cosmin Şandric, Ionuţ Cornel Mirea, Adrian Hăgătiş, and Erika Stanciu

Abstract The presence of wilderness areas in the Carpathian Mountains suggests that the Natural habitats that are part of this region still maintain their ecological functionality, but there are chances that some of these were not included into the network of protected Natural areas. There is thus an urgent need to find a method that can reliably identify them. True wilderness areas are very hard to find in the European landscape, due to strong anthropogenic impacts, which include intensive and extensive agriculture combined with other forms of resource exploitation, such as logging, overgrazing, mineral extraction, infrastructure projects, energy production and so on. The main ecological functionalities of these habitats can be permanently disrupted, thus losing the true meaning of what these areas stand for. A significant negative impact has been repeatedly recorded on populations of umbrella species, such as European brown bear *Ursus arctos* (Káre 1978), and other large carnivores. These species can be used as indicators for conservation actions, because of the large areas covered by their home ranges, which in turn help to protect almost all other species of interest in the region.

Keywords Romania • South Western Carpathians • Wilderness quality mapping • Wild areas

---

D.Ş. Măntoiu (✉) • M.C. Nistorescu • I.C. Mirea
EPC Consultanţă de mediu, Environmental consulting, Bucureşti, Romania
e-mail: dragos.mantoiu@gmail.com

I.C. Şandric
Faculty of Geography, Environmental Sciences, University of Bucharest, Bucureşti, Romania

A. Hăgătiş
WWF World Wide Fund for Nature, Braşov, Romania

E. Stanciu
ProPark Foundation for Protected Areas, Braşov, Romania

S.J. Carver, S. Fritz (eds.), *Mapping Wilderness*,
DOI 10.1007/978-94-017-7399-7_10

10.1 Introduction

The wilderness concept can be best linked to the Romanian legislation as a "wild area". The National and Nature Parks Law, 49/2011 defines it as a site that has not been affected by human intervention or the level of any such intervention is deemed insignificant. The act also states that the wild areas must be afforded the highest protection status available that benefits from no human intervention. However, the term only applies within the limits of existing national or nature parks, and refers to their internal zoning, which translates into special zones for conservation, or park core areas (PCA). So far no attempts have been conducted to delineate these wild areas and to put them under wider protection. Furthermore, as our results show, wild areas are not to be found only in National and Nature Parks but also in NATURA 2000 sites as well as in regions with no level of protection. As such this can be regarded as a case of *de jure* versus *de facto* wilderness.

The study of wilderness areas in Romania is still in its early stages. The first steps were taken by a non-governmental organization (WWF Romania, also known as the World Wide Fund for Nature) with the support of an environmental consultancy company, EPC Consultanță de mediu, to create a method of identification, based on the local conditions, in what is considered to be one of the most Natural areas in Romania: the South Western Carpathians. The selected study site has a core of protected Natural areas (scientific reserves, national parks, Natural monuments, nature reservations, nature parks, geoparks and NATURA 2000 sites), that cover more than half of its land area. Less Natural areas are found towards the lowlands, and are more or less subject to human development, and as such are considered to be a suitable boundary for the wilderness modeling environment.

The main geographic features of the site are compact mountain chains, separated by depressions and deep valleys (Fig. 10.1). In the local landscape classification, this sector is composed of parts belonging to different larger landforms, such as the Southern Carpathians, in the central and eastern part, the Banat Massif in the southern part, and the Poiana Ruscă Mountains to the north-west. One of the most compact regions of undisturbed land can be found in the central part of the site. It contains mountains that have a radial distribution of their ridges, e.g. the Godeanu Massif (Roșu 1980), and glaciated landscapes with high barrier effects for the surrounding depressions, sometimes with steep slopes that have drops of more than 1000 m. The best example is located in the northern part of the Retezat Massif which is a former PAN Parks site.

The diverse landscapes of the study area offer support for lush temperate forests and rich alpine meadows, with altitudes ranging from the lowest recorded in the Carpathians (under 100 m in the Danube Gorges), to 2509 m asl in the Retezat Massif. These are divided in two major biogeographic regions: Alpine and Continental, harboring an important number of protected species, including large carnivores, such as European brown bear (*Ursus arctos*), Wolf (*Canis lupus*), Eurasian lynx (*Lynx lynx*) and Wild cat (*Felis sylvestris*).

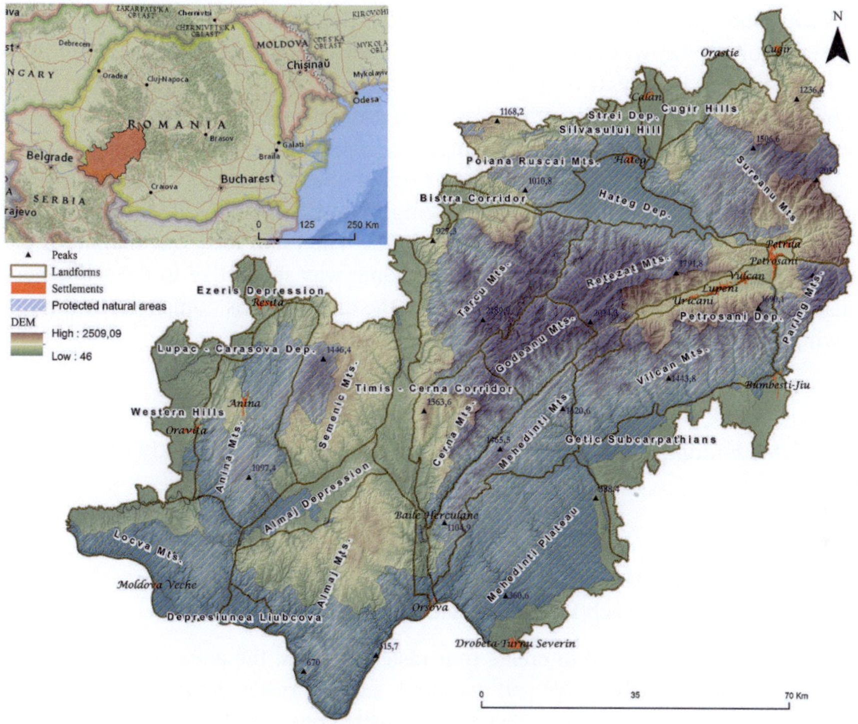

Fig. 10.1 Location of the study area and the main protected natural area network that includes: National and Nature Parks, Geoparks, and NATURA 2000 sites

10.2 Methods

The selected study area has long been seen as a landmark for an international protected Natural areas network before NATURA 2000 sites were designated. Pan Parks Foundation, which aimed to protect the last remaining wilderness areas in Europe, located one of their certified parks in the Retezat National Park, the first large protected Natural area in Romania which was established in 1935. It is from this core area that the whole concept of mapping a potential wilderness quality index has been developed, and then extended for the detection of smaller cores in the vicinity as well as potential corridors that may link them into a network. The goal is to detect which of the surrounding wilderness areas can be linked to one another, thus creating a potential wilderness network.

Mapping wilderness areas has proven to be a challenge. Variations can occur, by present or past use, having deep roots in the political options and choices for development of a country or a region. As most of the forests in the Carpathians have been exploited, one cannot assume that they will never become part of a wilderness area. If the forest is subject to Natural succession and reforestation, in time, it can reach maturity and maintain most of the original functionality. The methods for the detection of similar wilderness quality areas in different regions of the world have to be developed locally. It is crucial to only include features that have the greatest significance in the model, in other words, an accurate spatial representation of the potential or existing negative human impact.

The method used for this analysis was based on presumptions attributed to the spatial data obtained by "head-up" digitizing in an ESRI ArcGIS 10 environment, from the topographical maps and aerial images. The main idea was that all of the anthropogenic features were considered in some way restrictive to the wilderness quality index. The relevant data selected for the analysis was composed of the following items:

- Roads, with two types of attributes: paved and unpaved,
- Settlements,
- Isolated buildings, that include all of the constructions in the study site,
- Railroads,
- Power lines,
- Land use,
- Digital Elevation Model
- Statistical data for each local administrative unit, level 2 (LAU).

The features were used to create four raster datasets: the accessibility map, the isolated building density map, and two other support raster datasets, composed of statistical data at an LAU level. The datasets were considered a fair input for a spatial multi-criteria evaluation (Fritz et al. 2000), because they address, more or less, the main conservation issues in the area. The accessibility map, based on the types of roads, the digital elevation model and the land use, is a raster dataset that shows the time it takes to reach a pixel from the nearest road, in minutes. Different costs for human movement have been considered, combining horizontal costs, or the speed at which the average healthy and fit individual can move through different land use categories (Carver et al. 2008) and vertical costs, depending on the slope, obtained from the digital elevation model, with the same conditions applied to the individual, as mentioned earlier. The ease of movement on any road type was taken into account by a coefficient added to the vertical cost function (Tobler 1993). The isolated buildings density raster dataset was obtained by a deterministic interpolation (Sibson 1981) with values extracted from the centroid of a grid with the mesh size of one kilometer. The values represent the number of isolated buildings found in each grid. LAU level raster datasets were composed of indices that aim to describe the relevant human footprint in the area: the Natural growth of the population, that shows where aging trends are powerful and may lead to less impact on the environment, and an index developed by the Agency of Payment and Intervention in Agriculture (APIA 2011), describing the negative impact upon meadows and pastures generated by grazing animals. The total number of each animal category was transformed into adult cattle units, and the sum was divided by the total number of hectares for grazing space found in each LAU.

The accessibility map and the isolated building density map were used to give a spatial dimension of the human-influenced areas, eliminating portions of the Natural land use categories which could have suffered some form of alteration generated by human activities. Two main classes of human vital space were taken into account: the living space, that was covered by the results of the isolated buildings interpolation, which extended the area of influence more than a simple settlement boundary could have, at the same time giving a density value that can be more or less proportional to the relevant negative impact on the wilderness quality index and the resource gathering space, which was covered by the accessibility map. The main presumption was that most of the roads that were built in the Natural areas could have been used to gain access to any type of resource exploitation, and the values that were obtained near such elements were considered to have a low value wilderness quality index. The other two raster datasets obtained from statistical data were used to weight the influence of the accessibility map and the isolated building density map, in the idea that a thriving community will make better use of its available resources and infrastructure. The four raster datasets can be viewed in Fig. 10.2.

In order to obtain a single raster dataset that represents the wilderness quality of the study area, the four variables were first tested for correlation with ENM Tools

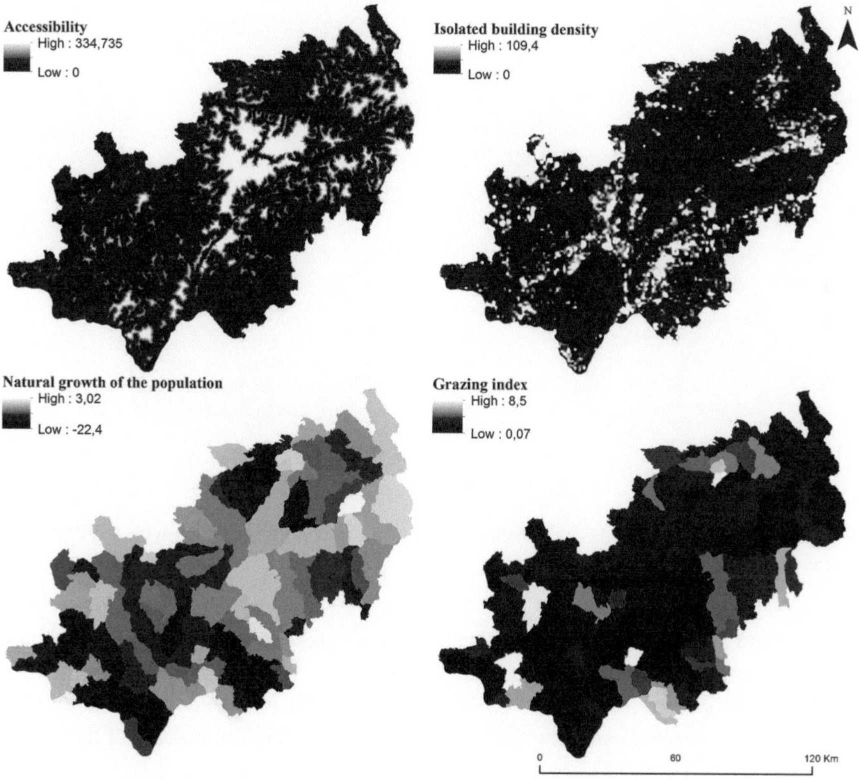

Fig. 10.2 Raster datasets used in the analysis were divided in two categories: the upper row contains the data that can isolate the human living space and the resource gathering space, while the lower row contains data that was used to amplify or decrease the influence of the first, at the LAU level

Table 10.1 Correlation matrix for the raster datasets used in the analysis

R value – Correlation matrix	Accessibility	Isolated buildings	Natural growth	Grazing
Accessibility	0	0.07	–0.03	–0.21
Isolated Buildings	0.07	0	–0.09	0.02
Natural Growth	–0.03	–0.09	0	0.03
Grazing	–0.21	0.02	0.03	0

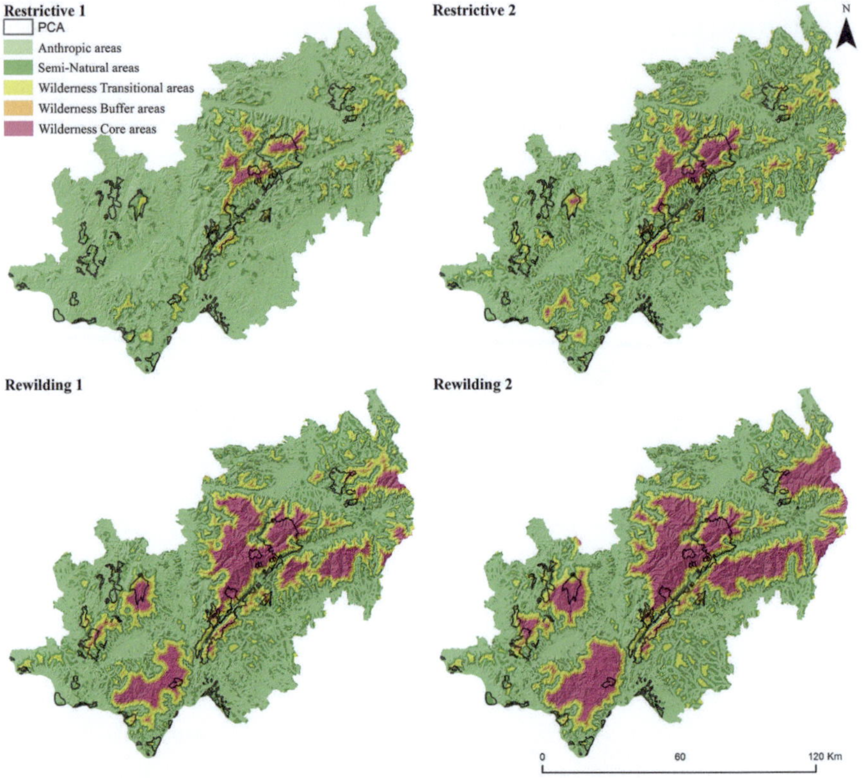

Fig. 10.3 Alternatives for the wilderness areas in the study site, compared with the Park Core Areas (PCA)

(Waren and Seifert 2011a, b, c, d). The results are presented in Table 10.1, and indicate no significant correlation between the variable. With the help of a perception study, weights were assigned to each of the variables in a weighted overlay tool. These were generated using the Saaty AHP scale, which reduces some of the bias in the analysis (Saaty 1977).

The results were reclassified by the Natural breaks classification into five intervals, which were coded as: anthropogenic, semi-Natural, wilderness transitional, buffer and core areas (Fig. 10.3). The model was run for four alternatives/scenarios, with modifications made to the raw data, presented in Table 10.2.

Table 10.2 Criteria taken into account for the wilderness alternatives

Wilderness alternatives					
No.	Input features	Restrictive 1	Restrictive 2	Rewilding 1	Rewilding 2
1	Roads	No	Yes	Yes	Yes
2	Isolated buildings	Under 3/km^2	Under 3/km^2	Under 5/km^2	Under 10/km^2
3	Settlements	No	No	Small isolated communities	Small isolated communities
4	Agricultural land	No	No	Yes	Yes
5	Power lines	No	No	Yes	Yes
6	Railroads	No	No	No	No
7	Grazing	No	Yes	Yes	Yes

The roads that were included in the more permissive models represent only isolated sectors of the mountain roads that do not end in an anthropogenic feature such as a clear cut or a chalet. Small isolated communities are represented by areas with a low density of houses, similar to the isolated buildings threshold but with a legal spatial establishment, such as the isolated villages from the Cerna Mountains.

The alternatives were created in order to give a spatial representation for the options that a rewilding process can bring, with the presumption that the negative impact which was generated until the mitigation measure in question was applied is not taken into account, and the affected areas will return to their Natural state in a certain amount of time.

10.3 Results

The datasets can reveal an interesting perspective upon the Natural areas in the SW Carpathians, but the limitations of the approach can be subject to debate. For example, the two indices mapped at an LAU level cannot explain much of the local variations, as they have well marked boundaries and only one value per unit, and therefore are subject to the modifiable areal unit problem (MAUP) (Openshaw 1984). The information should be sourced at a smaller scale in order to evaluate the wilderness quality index more accurately. The large cattle unit index cannot be assigned only to one LAU, because the herds usually graze on ridges covered with meadows and pastures, and these landforms are often used to mark the boundaries of an LAU, thus usually end up crossing them to find suitable grazing areas. In order to map this type of activity, a local survey must be created, defining the area of influence for each sheepfold and updating the number of animals in the herd.

Other uncertainties may be the subject of user accuracy and perception, mainly in the process of image interpretation, e.g. the roads used in the accessibility map can be incomplete, or some roads may have been temporarily or permanently abandoned by the forestry officials. This information, along with the age of the forest, species composition and exploitation plans, that include proposals for new forestry roads, should be provided by the Forest Research and Management Institute. Unfortunately the data is considered classified, because the site contains private forest districts. In order to increase the data credibility, a validation process was performed, with the help of custodians for each of the protected Natural areas that had some form of management e.g. all types of park administrations, NGOs and forestry officials. They have been provided with the raw data, and were encouraged to bring improvements based on their knowledge of the field.

At the same time, a survey was conducted to better understand the perception of the personnel on matters regarding wilderness quality, which was used to attribute weights to the final analysis. The survey had a total of 28 fields split into three main categories: the importance of elements that make up a wilderness area, the anthropogenic artifacts that have negative impact and the activities that pose a threat to wilderness, with answers split into ranks that ranged from 1, which meant a low negative impact, to 10. There were 11 respondents representing the different types of protected Natural areas, as follows: 5 national parks, 2 nature parks, 2 geoparks, 2 NATURA 2000 sites, SCI (Site of Community Importance). A multiple correspondence analysis was performed in order to compare the results of the last two categories (Fig. 10.4), which included 19 of the 28 fields that composed the survey. Results show spatial relations between the intensity of the problems shared by the protected Natural areas. The national parks form a cluster that differentiates them from the other classes, having similar responses for many of the fields, such as high negative impact of the mineral extraction activities, clear cuts, selective cuts, forestry roads, grazing, both forms of agriculture, and low negative impacts of the tourism activities. Subtle changes in values indicate some local problems, such as intensive speleological tourism in the parks that contain portions with karst landscapes.

The final results for the restrictive 1 alternative, that best resembles the sum of most of the requirements found in wilderness definitions worldwide (Landres et al. 2008), show five wilderness core areas with the highest class of protection according to the IUCN protected area management categories, found in the Retezat Massif. These include over 20 % of the identified core area in the Ia category, which translates at a national level, into scientific reserve (Table 10.3).

The rest of the alternatives gradually start to grow when restrictions are eliminated, initially forming a higher number of cores, but finally merging into six large areas. Two of them are divided by the DN66A national road, which for the moment has an unpaved portion with light traffic. Due to future road development planning, our model included the paved version of the road, which cuts through any wilderness core.

Fig. 10.4 (continued) *Eco* Ecotourism, *Spel* Speleological tourism, *SPt* SPA tourism, *ArchT* Archaeological tourism. Protected natural areas: **National parks**: RetezNP – Retezat, SemNP – Semenic, ChNerNP – Cheile Nerei – Beuşniţa, DomogNP – Domogled – Valea Cernei, DefJNP – Defileul Jiului, **Natural parks**: PortFierNATp – Porţile de Fier, GradMuncNATp – Grădiştea Muncelului – Cioclovina, **Geoparks**: MehGEOP – Platoul Mehedinţi, HategGEOP – "Ţara Haţegului", **NATURA 2000 sites, SCI**: NGorjSCI – Nordul Gorjului de Vest, TarcuSCI – Munţii Ţarcu

Fig. 10.4 MCA on the anthropogenic artifacts that have negative impact and the activities which pose a threat to wilderness areas. *Shp* Sheepfold, *Chal* Chalet, *Fr* Forestry road, *Pf* Planted forest, *Wr* Wood exploitation routes, *H* Hydro Energy, *Slo* slope fitting, *CC* clear cuts, *SC* selective cuts, *SubC* Subsistence cuts, *ExtAg* Extensive agriculture, *IngAg* Intensive agriculture, *ConP* Construction projects, *Min* Mineral extraction, *InfN* Infrastructure network, except forestry roads,

Table 10.3 Per cent of the total surface of the identified wilderness classes, split between the IUCN protected area management categories

Wilderness alternative – restrictive 1			Per cent of area occupied by IUCN protected area management categories					
No.	Name	Wilderness quality class	Ia	II	III	IV	V	No protection
1	Retezat	Transitional area	4.99	79.6	0	8.52	0	6.89
		Buffer area	10.24	81.76	0	3.99	0	4.01
		Core area	22.29	77.01	0	0	0	0.7
2	Ţarcu – Godeanu	Transitional area	0	53.2	0	45.84	0	0.96
		Buffer area	0	54.28	0	45.72	0	0
		Core area	0	39.29	0	60.71	0	0
3	Albele Ridge – Murgan Peak	Transitional area	0	0	0	99.41	0.59	0
		Buffer area	0	0	0	100	0	0
		Core area	0	0	0	100	0	0
4	Craiova Stream – Cerna Mts.	Transitional area	14.18	60.79	0	22.66	0	2.37
		Buffer area	15.91	84.09	0	0	0	0
		Core area	8.9	91.1	0	0	0	0
5	Parâng Mts.	Transitional area	0	0	0	97	0	3
		Buffer area	0	0	0	100	0	0
		Core area	0	0	0	100	0	0

Dudley (2008)

The main types of Natural ecosystems in the study area are grasslands and forests. A surface analysis was performed on these types of ecosystems, showing the per cent of each class of wilderness, grouped into each alternative (Fig. 10.4). The main conclusion is that the core area has a larger proportion of grassland if the selection criteria for the wilderness areas are more restrictive.

10.4 Conclusions

Designation of wilderness areas will never solve the true problems of conservation in the Natural protected areas of Romania; they simply represent the most inaccessible and intact portions of land that resulted from the interaction of human activities and the Natural areas. For now, wilderness areas should not be presented as a new proposal of new protected Natural areas, but should be considered to be an extension of the PCA in national or nature parks, that can benefit from no human intervention protection status. In the NATURA 2000 network the habitat conservation status should be evaluated and conservation measures should be proposed in a form of a management plan so that they will benefit from proper conservation actions, and the portions with no conservation status (i.e. those with de facto wilderness qualities) should be at least included into the NATURA 2000 network.

In another approach to map protected Natural areas or to try and concentrate the conservation funds towards sensitive areas, the results of this analysis could be merged with the sum of favorable sites obtained from species distribution models, or the high biodiversity value sites (Klein et al. 2009) which can be performed on the umbrella species, such as the large carnivores.

The method described can be seen as a first step in the identification process of wilderness areas in Romania. Even though the method is developed locally, the dataset that had the most impact upon the model, or the accessibility dataset, can be compared to the remoteness datasets used in other methods worldwide (Carver et al. 2002, 2012). The local development of the models should be in tandem with the methods used in other regions in order to have a common ground when comparing the final results.

Acknowledgments We would like to thank Erika Stanciu, the initiator of this project, for sure the main supporter of wilderness in Romania and also to Adrian Hăgătiş and Alexandra Doba who provided huge contribution to the project.

References

Carver, S., et al. (2002). Wilderness attribute mapping in the United Kingdom. *International Journal of Wilderness, 8*(1), 24–29.

Carver, S., et al. (2008). *Wilderness study in the Cairngorms national park, Final report*, University of Leeds, Leeds, UK.

Carver, S., et al. (2012). A GIS model for mapping spatial patterns and distribution of wild land in Scotland. *Landscape and Urban Planning, 104*, 395–409.

Dudley, N. (Ed.). (2008). *Guidelines for applying protected area management categories*. Gland: IUCN.

Fritz, S., et al. (2000). New GIS approaches to wild land mapping in Europe. *USDA Forest Services Proceedings RMRS-P, 2*, 120–127.

Káre, E. (1978). Human impact on a brown bear population (*Ursus arctos* L.). *Biological Conservation, 13*(2), 81–103.

Klein, C. J., Wilson, K. A., Watts, M., Stein, J., Carwardine, J., Mackey, B., & Possingham, H. P. (2009). Spatial conservation prioritization inclusive of wilderness quality: A case study of Australia's biodiversity. *Biological Conservation, 142*(7), 1282–1290.

Landres, P., et al. (2008). The matrix: A comparison of international wilderness laws. In C. F. Kormos (Ed.), *A handbook on international wilderness law and policy* (pp. 31–54). Golden: Fulcrum Publishing.

Openshaw, S. (1984). The modifiable areal unit problem, CATMOG 38. In Geo Abstracts, Norwich.

Roşu, A. (1980). *Geografia Fizică a României*. Bucureşti: Editura Didactică şi Pedagogică.

Saaty, T. L. (1977). A scaling method for priorities in hierarchical structures. *Journal of Mathematical Psychology, 15*, 57–68.

Sibson, R. (1981). A *brief description of natural neighbor interpolation, interpolation, multivariate data*, Chapter 2 (pp. 21–36). New York: Wiley.

Tobler, R. W. (1993). *Three presentations on geographical analysis and modelling* (Technical report, pp. 93–101). Santa Barbara: National Centre for Geographic Information and Analysis, University of California.

Waren, D. L., & Seifert, S. N. (2011a). Environmental niche modelling in Maxent: The importance of model complexity and the performance of model selection criteria. *Ecological Applications, 21*(2), 335–342.

Waren, D. L., & Seifert, S. N. (2011b). Agency for Payments and Interventions in Agriculture, Informative guide for farmers seeking support for agro-environmental measures in 2011, version 3.0, PPS-AG.

Waren, D. L., & Seifert, S. N. (2011c). Convention regarding the protection and sustainable development of the Carpathians (Conventia Carpatica, Kiev, 22 may, 2003), approved by the Law 389/2006.

Waren, D. L., & Seifert, S. N. (2011d). Government Emergency Decision No. 57 from June 2007, regarding the regime of Natural protected areas, conservation of Natural habitats, wild flora and fauna, approved by the Law 49/2011.

Chapter 11
Purism Scale Approach for Wilderness Mapping in Iceland

Rannveig Ólafsdóttir, Anna Dóra Sæþórsdóttir, and Micael Runnström

Abstract Coincident with increased utilization of the Icelandic highlands, its image as a unique and pristine wilderness is gradually changing. People's perception of wilderness is influenced by a number of factors relating to their culture and socio-economic background. Furthermore, how people value pristine land or define wilderness varies depending on the location and function of the assessment. Therefore, understanding perceived wilderness is likewise of major importance in the planning and long term management of tourism within the Icelandic highlands. This paper attempts to identify and map perceived wilderness areas within the southern Icelandic highlands, using the purism scale approach. The results indicate that constructions related to power plants (i.e. plants, power lines, and dams) are considered undesirable by all four tourism market groups. The results moreover show that non-purists visiting the Icelandic highlands do not favour paved roads. Conversely, mountain huts do not affect the perceived wilderness for any of the purism groups. The perceived wilderness mapping of the southern Icelandic highlands shows that nearly the whole area, or 97.2 %, is perceived as wilderness by the non-purism group, while less than half, or 45.4 %, is perceived as wilderness by the strong purism group. Once a wilderness area becomes known as a tourist destination, maintaining its wilderness condition becomes increasingly difficult. In order to avoid the overuse of wilderness for tourism and other economic sectors, ambitious planning and appropriate management are critical. This includes identifying limits of growth and further development. Without such limitations, the use of wilderness is simply unsustainable.

Keywords Wilderness mapping • Tourist perception • Purism scale • Tourism • Iceland

R. Ólafsdóttir (✉) • A.D. Sæþórsdóttir
Department of Geography and Tourism, Faculty of Life and Environmental Sciences, University of Iceland, Sturlugata 7, IS-101, Reykjavík, Iceland
e-mail: ranny@hi.is

M. Runnström
GIS Centre, Department of Physical Geography and Ecosystem Science, Lund University, Sölvegatan 10, SE-223 52, Lund, Sweden

S.J. Carver, S. Fritz (eds.), *Mapping Wilderness*,
DOI 10.1007/978-94-017-7399-7_11

11.1 Introduction

> Travelling alone on foot in this vast and threatening landscape was one of the most incredible and spiritual experiences of my life. (Miriam Rose 2006).

Still, the Icelandic wilderness areas provide unique experience for tourists. The Icelandic wilderness resource has however witnessed a rapid expansion of natural resource exploitation that still seems to be progressing. Spectacular nature and vast wilderness have long been the predominant attractions for tourists visiting Iceland. Consequently, they form the backbone of the growing tourism industry that is currently the nation's second largest export sector (Statistics Iceland 2013a). Icelanders numbered 321,857 on the 1st of January 2013 (Statistics Iceland 2013b). This small population shares a landmass of approximately 103,000 km^2 and has throughout the 1100 years of the island's human settlement been distributed mainly along the coastline, leaving the interior highlands a largely uninhabited wilderness (e.g. Ólafsdóttir and Runnström 2011a; Sæþórsdóttir et al. 2011). Previously the Icelandic highlands were used only for summer grazing, but from the early 1970s onwards, gradual changes towards multiple uses have taken place. Thus, vehicles have taken over the role traditionally played by horses in the rounding up of the sheep in autumn, numerous hydro-electric power stations have been constructed and tourism is growing rapidly. This increased use of the highlands and the consequently increased demand for motorized vehicle access into the highlands has led to shrinkage of the country's wilderness area (e.g. Ólafsdóttir and Runnström 2011a; Sæþórsdóttir et al. 2011).

For centuries the Icelandic interior highlands were of little economic significance; they were considered poor pastures and presented a substantial obstacle when travelling between different parts of the country (e.g. Sæþórsdóttir et al. 2011). This, together with the country's sparse population which fluctuated between 30 and 50 thousand until the nineteenth century when the population began a steady increase (e.g. Karlsson 1975; Karlsson and Kjartansson 1994; Júlíusson 1995), preserved the Icelandic highlands from human impact until relatively recently. Until World War II, the only access into the highlands was by foot or horse, but when the British-American occupation forces imported all-wheel-drive army trucks they opened up the vast wilderness of the highlands, as these vehicles were able to traverse the large glacier rivers and drive through rough terrain. Using 4 × 4 vehicles, the classical 'highland safari' was developed which became the first significant form of organized tourism in the Icelandic highlands (Huijbens and Benediktsson 2007; Sæþórsdóttir et al. 2011). In the 1930s, the Iceland Touring Association had begun building mountain huts in the interior highlands for recreational purposes (e.g. Ólafsdóttir and Runnström 2013). At the present time, there are hundreds of mountain huts in the Icelandic highlands, most of them unlicensed (Ministry of the Environment and The National Planning Agency 1999). In the 1970s the development of large-scale power production began in the highlands when the first hydro-electrical power plant was constructed in the southern periphery of the area (e.g. Pálsdóttir 2005; Sæþórsdóttir 2012a). Over time, additional plants have been constructed and older ones enlarged. Today, seven power plants are located within the highlands, most of them in the southern highlands (Landsvirkjun 2013). The construction of power plants led to improved access into the highlands as roads were constructed and rivers bridged. As the highland road network grew and road conditions improved, day-tripping into the highlands became easier. Yet, to a large extent, the highland roads are rough gravel roads or tracks passable only by 4×4 vehicles. However, concurrent with increased tourism, the Icelandic highlands are growing in popularity with increased demands for improved infrastructure. More than one third (36.3 %) of all foreign summer tourists visit the interior highlands (ITB 2012). Thus, the highlands and their wilderness are a valuable resource to the Icelandic tourism industry, as well as being of symbolic value used in various visual media e.g. used as a major marketing slogan in the symbolic economy.

Recent mapping of Icelandic designated wilderness areas (i.e. Ólafsdóttir and Runnström 2011a, b; Taylor 2011) indicates that the area free from man-made structures currently covers about one third of the total surface area of the country, and that Iceland has lost up to 70 % of its wilderness areas since the 1930s. Likewise, road-less areas have gradually decreased, with a consequent decrease in the quality of the wilderness. The increased number of land use conflicts stemming from the more intense utilization of the highlands (e.g. Thórhallsdóttir 2007; Benediktsson 2008; Sæþórsdóttir 2012a) underline the importance of forming a better knowledge and understanding of Icelandic wilderness resources, in order to implement sustainable management of the country's remaining wilderness areas. This is particularly true in terms of government aims of sustainable development (*cf.* Ministry of the Environment 2010) as well as for the organization of sustainable tourism in the Icelandic highlands. However, despite the rapid increase in human interference and the consequently changed appearance of the Icelandic highlands, research reveals that many travellers still experience the area to be wild and unspoiled nature (Sæþórsdóttir 2010b). Therefore, understanding perceived wilderness is also of major importance for the long term planning and management of tourism within the Icelandic highlands. This paper attempts to identify and map perceived wilderness within the southern Icelandic highlands, using the purism scale approach.

11.2 Previous Mapping of Icelandic Designated Wilderness Areas

The first step towards designating Icelandic wilderness areas was taken in 1997, following a governmental decision concerning a strategy for the preservation of pristine wilderness in Iceland. Subsequently, a work group was appointed by the Icelandic Minister of the Environment which formulated an official definition of Icelandic wilderness. This definition reflects conventional definitions, corresponding to the original US Wilderness Act of 1964, defining Icelandic wilderness to be an area of land:

- where no trace of human activity is to be found and the natural landscape develops outside of any pressure related to human influence.
- which is situated a minimum distance of 5 km from any human structure or infrastructure, such as roads, houses, power lines, telecommunication masts, dams, etc.
- which is at least 25 km^2 in size, or of such a size that one may enjoy solitude and the natural landscape without disturbance from human structures or traffic from mechanized vehicles.

Icelandic Act no. 44/1999 on Nature Conservation, section 3 (authors' translation)

Recently, a new act on nature conservation, no. 60/2013, has been issued by the Icelandic parliament and is expected to come into effect in 2015. With regard to wilderness, the new act still embraces the same definition as the previous act, no. 44/1999, however, the new act contains additional categories of protection, among which one is aimed at uninhabited wilderness (i: óbyggð víðerni) which will become a legal status of protection. The official mapping of designated wilderness areas in Iceland still only takes into account the criteria of 5 km distances from major roads (Fig. 11.1).

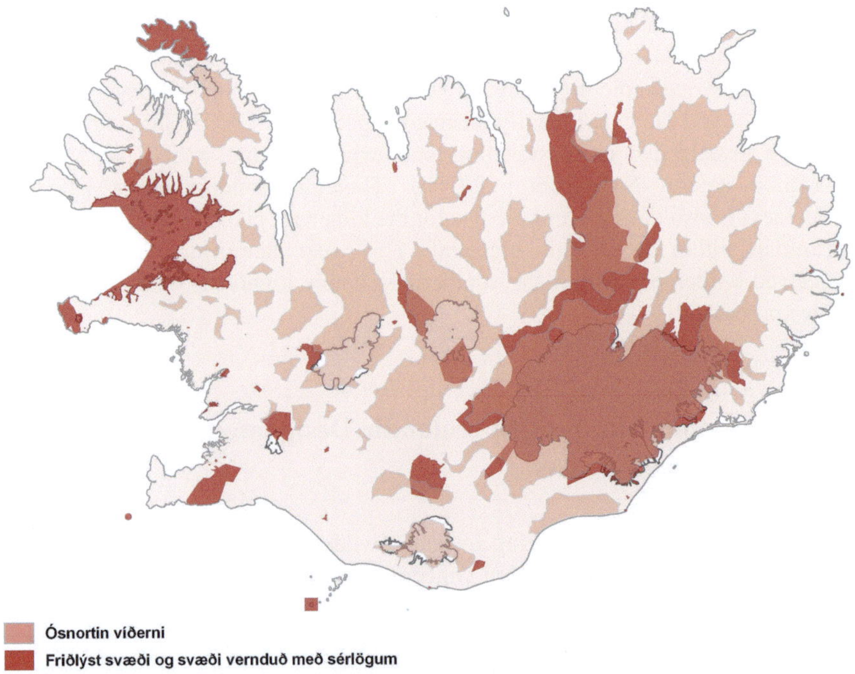

Fig. 11.1 Official mapping of Icelandic wilderness. *Pink* signifies *pristine wilderness* and *red protected areas* (The Environment Agency of Iceland and National Land Survey of Iceland 2009)

The first comprehensive assessment of Icelandic wilderness was carried out by Ólafsdóttir and Runnström (2011a). Their assessment is based on the official definition of the wilderness concept, as stated in the Icelandic act no. 44/1999, applying two different methods; firstly, a proximity analysis, where buffer zones were created and categorized, based on similar criteria as previous studies (e.g. Lesslie and Taylor 1985); and secondly, a viewshed analysis, where what is actually visible in the landscape in relation to topography is taken into account. Both these analyses are based on geographical digital data on a national scale, obtained from the National Land Survey of Iceland.

The proximity analysis mapping is based on three factors: remoteness from mechanized access, remoteness from settlement and apparent naturalness. Each factor was categorized into different attribute variables and appropriate disturbance distances were determined for each variable (*cf.* Ólafsdóttir and Runnström 2011a). Maps of all three factors were combined in a geographical model to obtain a holistic map demonstrating the total disturbance distances from all attribute variables used (Fig. 11.2). According to the proximity analysis results, the area outside the integrated buffer zones makes up 34,695 km^2, or 34 % of the surface of Iceland. Out of these 34 % the country's ice caps cover 26 %. Taylor (2011) added a temporal factor to the proximity analysis by assessing the change in areas free from roads and power lines between 1936 and 2010. Her results indicate that the number of polygons larger than 200 km^2 in size has decreased by over 70 % over the course of those

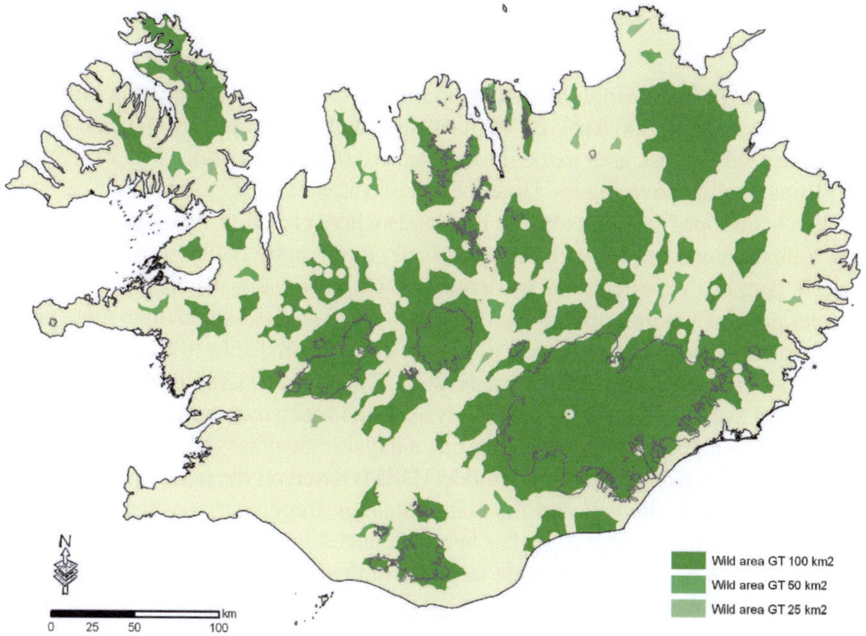

Fig. 11.2 Extent of Icelandic wilderness based on proximity analysis (From Ólafsdóttir and Runnström 2011a)

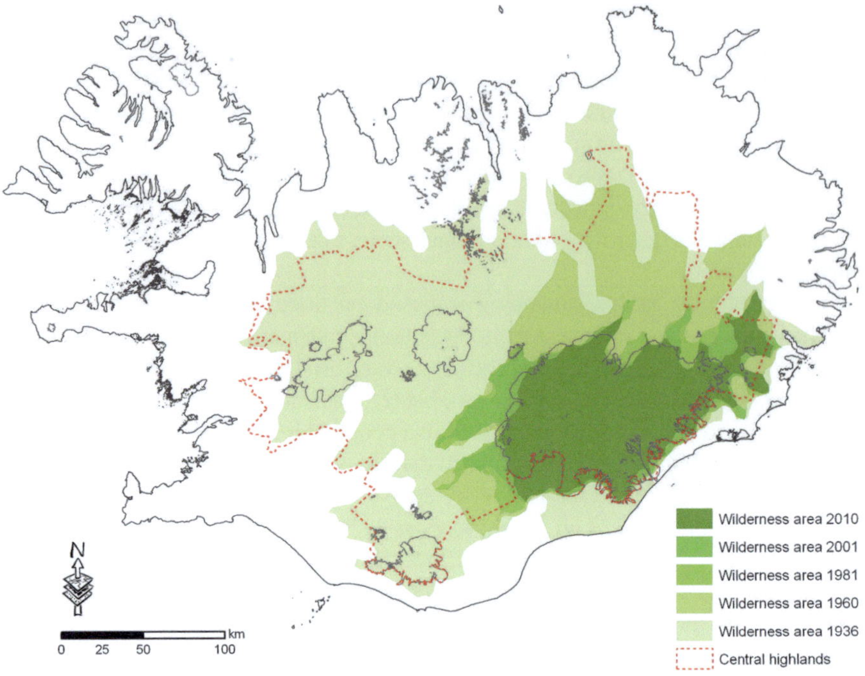

Fig. 11.3 Temporal decrease in the largest area free from roads and power lines within the Icelandic highlands (Modified from Taylor 2011)

75 years. In 1936 the single largest wilderness area within the Icelandic highlands made up nearly 50,000 km^2, or 47 % of the total land area. In 2010 the largest remaining wilderness area made up less than 10,000 km^2, or only 9 % of the total land area, mainly covering the large ice cap of Vatnajökull (Fig. 11.3).

In a landscape like the Icelandic highlands where elevation varies greatly, creating a mountainous and undulating landscape, topography is likely to play a large role in people's experience of wilderness. What is actually visible from different locales may be a more important variable in wilderness assessment than mere proximity to anthropogenic structures, as features may be invisible from certain angles and locations even though the distance is short. Therefore Ólafsdóttir and Runnström (2011a, b) also applied a viewshed analysis to the assessment of the Icelandic wilderness. The algorithm underpinning this analysis calculates the vertical angle for each grid cell in a digital elevation model (DEM) based on the relative difference in elevation between the cell and the cell containing the object and their horizontal distance. After the vertical angle has been calculated for each grid cell, the program compares each cell's vertical angle stepwise in the lines of sight, starting from the cell containing the object. If the vertical angle for a cell is lower than all cells closer to the object in the sight line, the cell is coded visible. However if the vertical angle

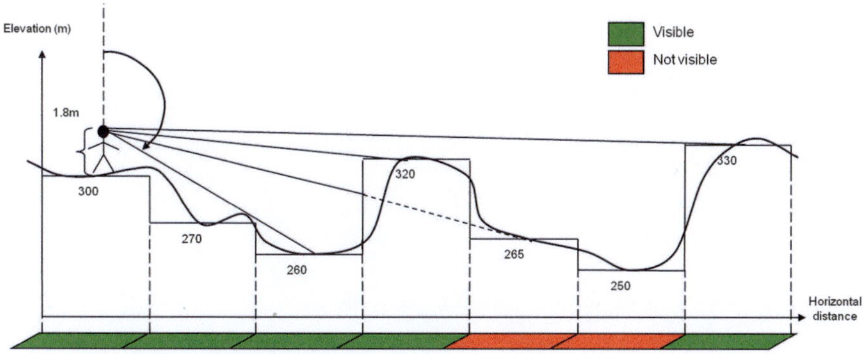

Fig. 11.4 Topographical impact on line of sight. The *numbers* represent elevation value in each grid cell in the DEM (From Ólafsdóttir and Runnström 2011b)

is higher than any one cell closer to the object it is coded as invisible (Fig. 11.4). What is visible is furthermore affected by the maximum sight distance, which is limited by the curvature of the Earth. Ólafsdóttir and Runnström (2011a, b) calculated the maximum sight distance (d) by using Pythagoras' theorem (Eq. 11.1), which shows that a person standing on a level expanse with eyes 1.8 m above the surface level is a 4.8 km distance from the horizon.

$$d = \sqrt{2Rh + h^2} \tag{11.1}$$

where d is the maximum sight distance; R is the Earth's radius (6,371,000 m); and h is the height of the viewer's eyes above ground level. However, if the person stands on a hill, the distance to the horizon is greater. Similarly, if the viewed object is tall, e.g. a power-line tower, it can be seen from further afield. As an example, an object of 2 m height, such as a car, will according to eq. 1 be invisible from about a distance of about 10 km (i.e. ~4.8 km from viewer's eyes to the horizon + ~5.0 km from the horizon to the car's roof). Therefore, as most of the anthropogenic objects in the Icelandic highlands are still rather small in relation to the topography of the landscape and, furthermore, are not striking in colour, a 10 km maximum sight distance was used in the viewshed analysis, at which distance objects were assumed to be too far away for disturbing visualization. The viewshed analysis was run for each of the anthropogenic features used in the proximity analysis model, in order to compare the outcomes of the two methods applied. On a national scale, the map resulting from the viewshed analysis shows a wilderness pattern and areal coverage similar to the one obtained from the proximity analysis (Fig. 11.5). On a local scale, however, a much more dynamic pattern emerges, which is closely interrelated to landscape topography.

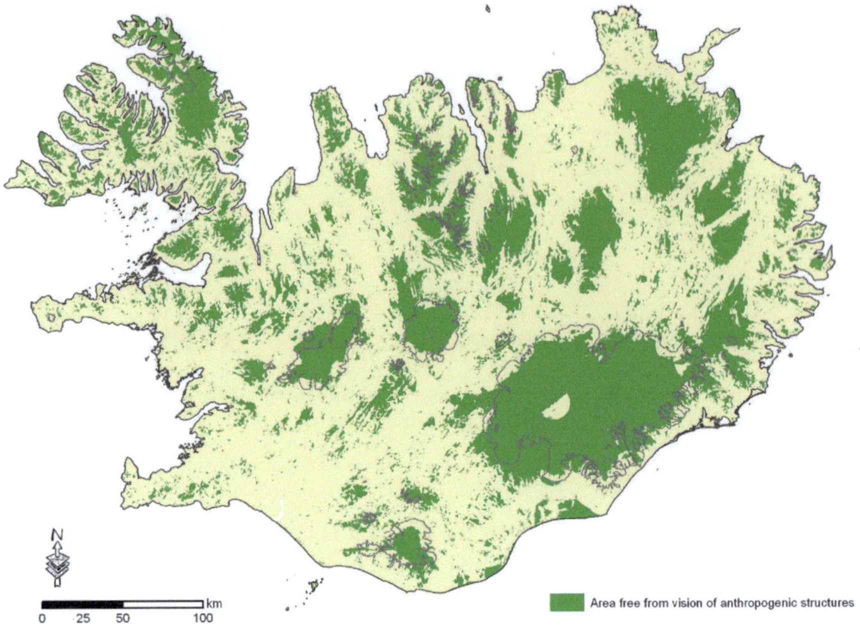

Fig. 11.5 Extent of Icelandic wilderness based on viewshed analysis (From Ólafsdóttir and Runnström 2011a)

11.3 Perceived Wilderness Mapping Based on Purism Scale Approach

11.3.1 Visitors' Perception of Wilderness

The overall perception of the Icelandic environment, with regards to tourism, seems largely based on a romanticized notion of its uniqueness and pristine wilderness (e.g. Ísleifsson 1996, 2009; Gössling and Hultman 2006; Oslund 2011; Sæþórsdóttir et al. 2011), an image which the Icelandic tourist industry is enthusiastic to maintain. However, concurrent with the increased utilization of Icelandic wilderness resources, this image is gradually changing. Hence, Taylor (2011) stresses that if Iceland does not maintain and manage its wilderness, overuse and overcrowding of the most popular areas might lead to dissatisfaction of tourists, decreasing the probability of their returning and adversely affecting the image of Iceland's wilderness as a tourist destination. Importantly, individual wilderness perception may contrast with the official definitions of wilderness. A number of studies have been undertaken to test people's individual perceptions of wilderness and their reasons for visiting such areas (e.g. Kliskey and Kearsley 1993; Higham 1998; Higham et al. 2001; Carver et al. 2002; Sæþórsdóttir 2010a, b; Van den Berg and Koole 2006; Flanagan and Anderson 2008; Lupp et al. 2011). The perception of wilderness by individuals is influenced by a number of factors relating to their culture and their socio-economic background, including age, gender, and education level (e.g. Sæþórsdóttir and Stefánsson 2009; Lupp et al. 2011). Furthermore, how people value pristine land or define wilderness varies depending on the location and the function of the assessment. According to Stankey and Schreyer (1987), the most common reasons for visiting wilderness areas are to experience solitude and unspoilt nature, as well as to escape from urban lifestyle.

Worldwide wilderness areas seem to be growing in popularity with all types of tourists, including so called "urbanists" who, although they are motivated to experience wildernesses, also require more facilities and services than their purist counterparts who prefer to have few or no facilities and to experience nature in an unspoilt environment (e.g. Sæþórsdóttir 2013; Taylor 2011). Thus, the increasing popularity is often met by expanding infrastructure to meet the increased demands. In this regard Sæþórsdóttir (2008, 2010b, 2013) points out that the increasing popularity and consequently increased crowding of many wilderness areas in the Icelandic highlands negatively impacts the expected wilderness experience, causing areas to become less attractive to the purist tourists, causing displacement of tourism to other, previously undisturbed, isolated areas. Hence, in this sense, the idea of wilderness is socially constructed (e.g. Williams 2002) and ever changing. In this subjective sense, wilderness does not exist without an observer to experience it and is more of an idea than an ontological phenomenon (i.e. Cronon 1998; Tuan 1990; Williams 2002). This is underlined by Van den Berg and Koole (2006), who point out that if people are unaware of previous human interference in an area, this interference does not detract from their wilderness experience. Thus, mapping perceived wilderness is critical for planning and managing wilderness tourism in the Icelandic highlands in a sustainable manner.

11.3.2 The Purism Scale

The purism scale is a continuum that ranks individuals in terms of their level of ideological attachment to purity or primitiveness, in their perception of wilderness (Fig. 11.6). Many variables, such as the level of infrastructure and available services and the density of tourists, influence the individual's perception and experience when visiting wilderness areas, based on his/her background and interests. These

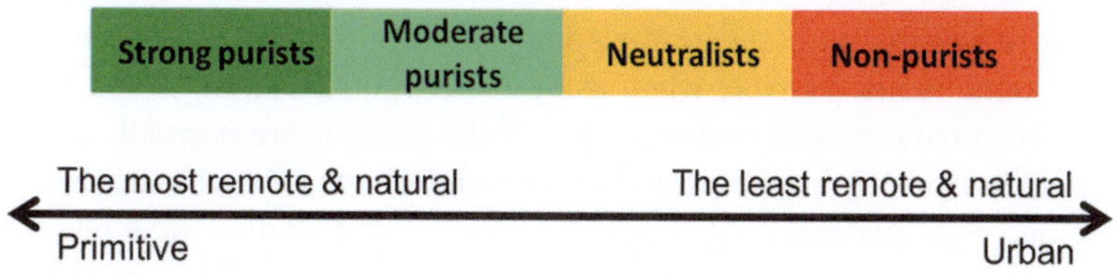

Fig. 11.6 The purism scale

variables reflect different needs, attitudes and expectations as well as the diverse tolerances of different types of individuals towards human impact on the environment. Thus, to distinguish the various types of wilderness perception, individuals with similar responses are grouped together, i.e. in "purism groups" (Hendee et al. 1968; Stankey 1973). In this regard Sæþórsdóttir (2010a) points out that some tourists are not sensitive to human-induced changes, whether they are buildings, roads or information signs. Conversely, such changes can ruin the experience of nature for those who enjoy a natural environment only if it is totally free of human alteration. Therefore, tourists have different opinions about what facilities and services are desirable, and it is obviously not possible to please everyone at a single location. For a natural destination, such as the Icelandic highlands, it is necessary in order to be competitive to distinguish the market segments when developing a natural tourist area, i.e. one has to ask and analyse; what type of tourist does this area attract and who could it potentially attract. The advantage of distinguishing market segments in this way is that neither time nor money is spent trying to attract tourists to a place in which they have no interest, or which they do not appreciate (e.g. Mohsin 2005; Buhalis 2000).

As thoroughly reviewed by Sæþórsdóttir (2010a), Hendee et al. (1968) were the first to analyse the different attitudes held by tourists towards wilderness areas within the USA, and, based on this analysis, suggest how tourism in wilderness areas should be managed. Based on their results, they categorized visitors into five groups on a so called Wildernism-Urbanism Scale, i.e.: *strong wildernists, moderate wildernists, weak wildernists, neutralists and urbanists*. They conclude that wildernists are more sensitive than the other visitor groups in their perception of the wilderness and its qualities, as defined in the US Wilderness Act, such as solitude and primitiveness. In 1973 Stankey carried out similar research, also in the USA. Based on visitors' responses to 14 items, he categorized them into four groups, i.e.: *strong purists, moderate purists, neutralists, and non-purists*, which he located on a so-called Purist Scale. Based on similar criteria, Schreyer (1976) produced another scale, the Wilderness Purism Scale, using 17 items to categorize visitors according to their attitudes. Wallsten (1988) was the first to apply this method in Scandinavia. He uses Stankey's (1973) terminology but his method of categorizing visitors differs from that of Stankey. While Stankey set fixed limits to distinguish between groups, Wallsten uses the Normal distribution. Vistad (1995) and Fredman and Emmelin (2001) use the same approach as Wallsten in assessing wilderness areas within the Scandinavian mountains. Sæþórsdóttir (2010a) points out that the different approaches taken by the Scandinavian and US scientists may have yielded somewhat different results. When fixed limits are used for the different categories, it is possible to compare the composition of visitors in different regions, while the same is not possible when using Normal distribution, as the limits will differ between datasets and research areas. If the datasets are normally distributed, different areas can only be compared by converting all results to the standard normal distribution. The advantage of using the Normal distribution method is that it highlights the differences between visitors at each location. This is useful when looking at certain locations and how to plan them, based on the requirements of different types of tourists. In Iceland Sæþórsdóttir (2010a, b, 2011) carried out a study using fixed limits to divide the different groups, as her main purpose was to compare user groups at different natural destinations in Iceland. Her results show that tourists with puristic attitudes constitute the majority of visitors in the least developed and least accessible tourist destination in the Icelandic highlands. Urbanistic views, on the other hand, are most common among visitors to the national parks and nature destinations in the lowlands. Travellers in the interior highlands generally want less development and fewer services than travellers in the lowlands and they are satisfied with the existing primitive conditions. Their satisfaction does not increase with more infrastructure and services; on the contrary, they prefer to travel in as natural an environment as possible.

Accessibility, physical environment, facilities and services are the critical factors determining which purism group will visit each area and these factors account for the different composition of visitors. Furthermore, the more popular the Icelandic highland areas become, the more likely it is that the composition of travellers will change. Increased numbers of visitors will drive away those who are most sensitive to crowding, and more visitors require more infrastructure. A new market group makes more demands on goods and services as can already be seen in Landmannalaugar, which has become the most visited destination in the Highlands, with over one hundred thousand tourists visiting the area annually. As a result, the character of tourism there has changed and, according to one third of visitors, there are too many tourists in the area (Sæþórsdóttir 2013).

11.3.3 Mapping Perceived Wilderness Using the Purism Scale Approach

In order to map the wilderness perception of tourists in the Icelandic highlands it was decided to use the methodology introduced by Kliskey and Kearsley (1993), Higham et al. 2001, and Flanagan and Anderson (2008) who base their mapping on existing data from questionnaires focusing on visitors' perception of different anthropogenic structures. This research builds on questionnaire surveys gathered among travellers at seven destinations in the southern Icelandic highlands from 2007 to 2011 (Fig. 11.7; Table 11.1). Completed questionnaires were received from 3288 visitors, with a response rate between 70 and 95 % (Sæþórsdóttir 2012b). The data was processed according to the purism scale approach (Fig. 11.8), using the score range from Sæþórsdóttir (2010a) to define each purism group in the Icelandic highlands as follows: strong purists scored >60; moderate purists scored between 50 and 59; neutralists between 40 and 49; and non-purists had scores of<40. The questionnaire was composed of 39 questions, only some of which are considered in this paper. These focus on the respondents' opinions on the desirability of various facilities and structures (e.g. paved roads, accommodation, power plants, etc.) at the location where the questionnaire took place. All questions were presented as a 5-point Likert scale, except three which were statements. To determine from the

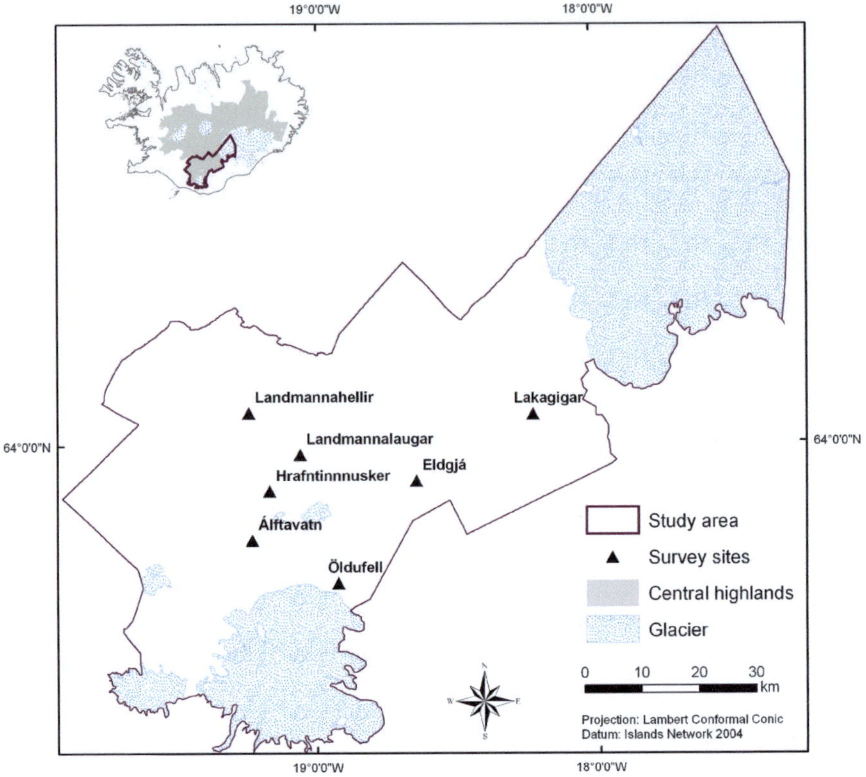

Fig. 11.7 The study area for the perception of wilderness mapping within the Icelandic southern highlands and the location of the seven questionnaire survey collection points

Table 11.1 Data used for the wilderness perception mapping

ID	Tourist destination	Year of survey	Questionnaires (n)	Per cent (%)
1	Landmannahellir	2011	180	5
2	Hrafntinnusker	2011	351	11
3	Álftavatn	2011	219	7
4	Eldgjá	2011	437	13
5	Öldufell	2011	58	2
6	Landmannalaugar	2009	1646	50
7	Lakagigar	2007	397	12
Total			**3288**	**100**

Obtained from Sæþórsdóttir (2012b)

data which features are considered undesirable in the wilderness area, the average score is calculated for each purism group. When the calculated average is below 3, this is mapped as an indication that respondents are against the current facility/ structure. Likewise, when the calculated average is above 3, it is assumed that

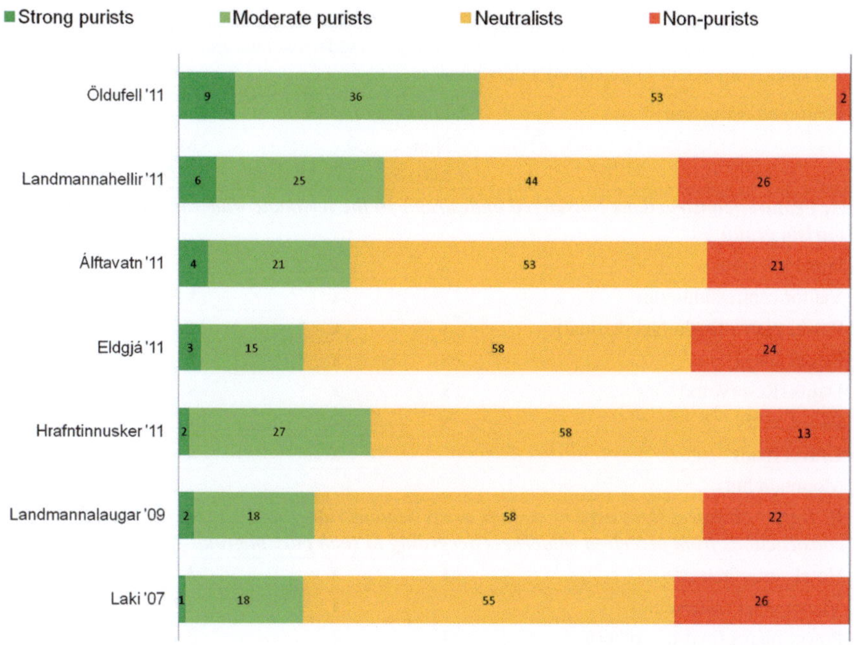

Fig. 11.8 Division of visitors according to the purism scale (Based on data from Sæþórsdóttir 2012b)

respondents are positive towards the facility/construction. The three statements-questions all focus on the visitors' opinions on power lines, dams and reservoirs, e.g. "Power lines may be present in an area which is considered wilderness". If over 50 % of respondents reply yes, this is mapped as if the construction is accepted.

The results from the perception analysis indicate that constructions related to power plants (i.e. plants, power lines, and dams) are considered undesirable by all four purism groups (Table 11.2A). This contradicts the results of Flanagan and Anderson (2008), which indicate that all features are accepted in the wilderness setting by the non-purism group. The results moreover show that non-purists visiting the Icelandic highlands do not favour paved roads. On the other hand, mountain huts do not affect the perceived wilderness by any of the purism groups (*cf.* Table 11.2A). In order to be able to map the different perceptions, an approximated distance of tolerance is required for each purism group (e.g. Higham et al. 2001; Flanagan and Anderson 2008). As such figures can be difficult for a tourist to define, and as such figures were not set forward in the questionnaires used, estimated numbers are given based on Higham et al. (2001)) and Flanagan and Anderson (2008), where the different buffer distances are supposed to reflect the graduated intensity of wilderness feeling by the four purism groups. The buffer distances are increased by one km per purism group (Table 11.2B). Several desirable wilderness features are not taken into account at this stage, due to lack of field data and/or lack of digital data (Table 11.2C).

Table 11.2 Features and structures considered undesirable in the Icelandic wilderness setting divided by purism group, and estimated buffer distances used to exclude areas featuring undesirable structures from extent of perceived wilderness

Feature/construction		Purism group			
		Non-purists	Neutralists	Moderate purists	Strong purists
A: Features/constructions considered undesirable in the Icelandic wilderness setting by purism group					
Hotel/guesthouse		x	x	x	x
Visitor centres/museum			x	x	x
Power plants (hydro/geothermal)		x	x	x	x
Power lines		x	x	x	x
Dams (Reservoirs)*		x	x	x	x
Paved roads		x	x	x	x
Gravel roads			x	x	x
Mountain huts					
B: Buffer distances (km) used to exclude areas demonstrating undesirable features/constructions from perceived wilderness according to each purism group					
Hotels/guesthouses		**1**	**2**	**3**	**4**
Visitor centers/museums			**1**	**2**	**3**
Power plants (hydro/geothermal)		**1**	**2**	**3**	**4**
Power lines		**1**	**2**	**3**	**4**
Dams (Reservoirs)[a]		**1**	**2**	**3**	**4**
Paved roads		**1**	**2**	**3**	**4**
Gravel roads			**1**	**2**	**3**
Mountain huts					
Farms					**4**
C: Lack of data					
Evidence of off-road driving	LDD				
Marked hiking routes	LDD				x
Designed footpaths	LDD				x
Tracks (dirt roads)	LFD				
Signposts/information signs	LFD				
Radio/telephone mast	LFD				
Maintained campsites	LFD				
Toilet facilities	LFD				
Commercial recreation (e.g. guided tours)	LFD				
LDD = Lack of digital data					
LFD = Lack of field data					

[a]In the questionnaires, respondents were only asked about their perception to dams

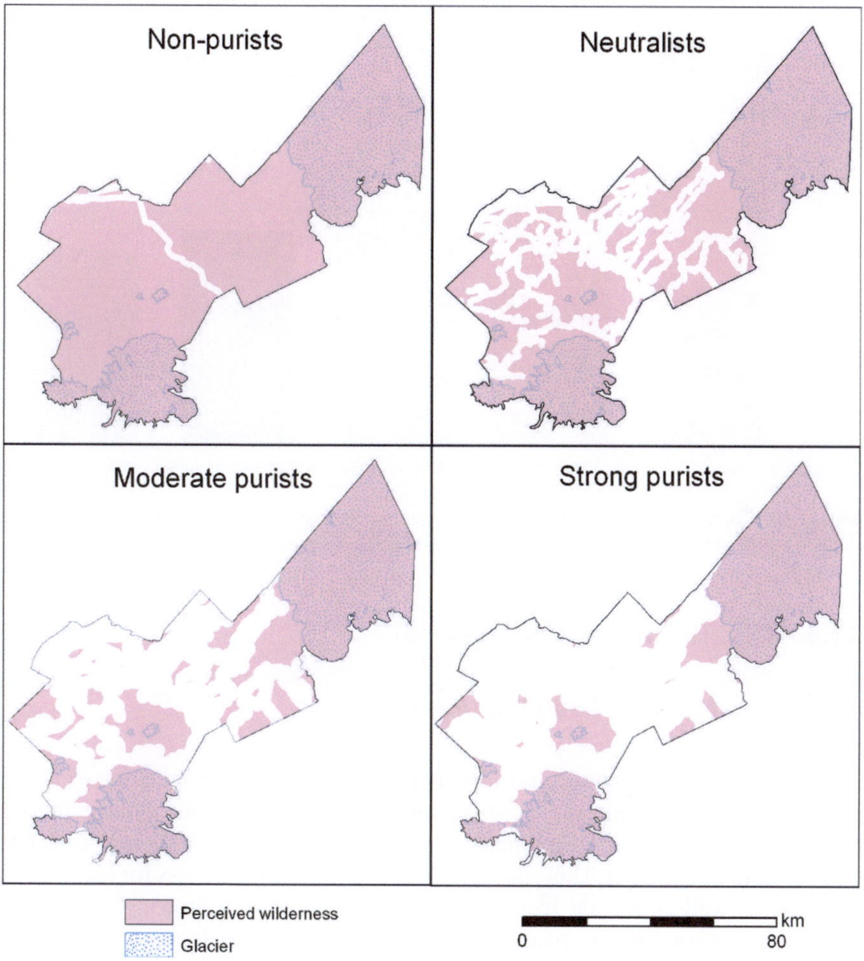

Fig. 11.9 The spatial and areal difference of perceived wilderness in the study area, using the criteria expressed by the four purist groups, i.e. non-purists, neutralists, moderate purists, strong purists

The resulting mapping of the perceived wilderness for the southern Icelandic highlands show that nearly the whole area, or 97.2 %, is perceived as wilderness by the non-purism group, whereas less than half, 45.4 %, is perceived as wilderness by the strong purism group (Fig. 11.9). As a point of comparison, designated wilderness areas (i.e. according to the proximity analysis) make up 49,7 % of the study area (Fig. 11.10).

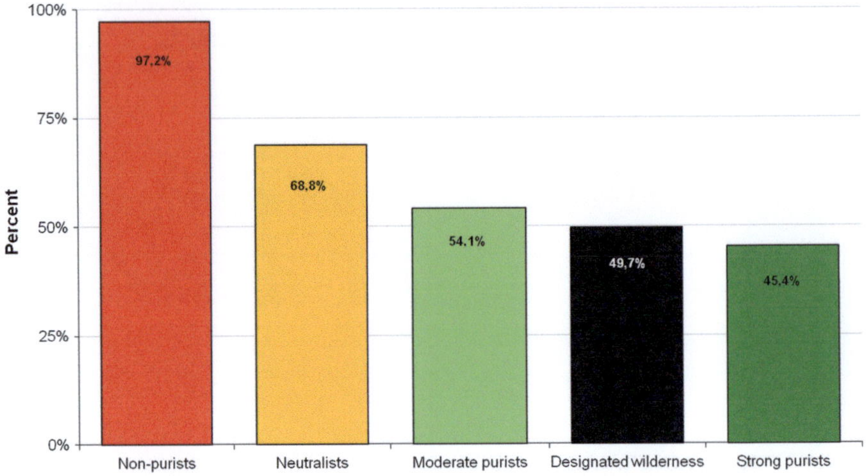

Fig. 11.10 Comparisons of the extent of wilderness as perceived by the four purism groups and as defined by the official designation of Icelandic wilderness (percentage of the study area)

11.4 Discussion and Conclusions

11.4.1 Wilderness Mapping in Iceland

The image of pristine nature has long been employed when marketing Icelandic exports. Likewise, the Icelandic tourism industry has long been using the country's wilderness as their main selling point when attracting tourists (Sæþórsdóttir 2008; Sæþórsdóttir et al. 2011). Thus, the value of the Icelandic wilderness for the tourism industry, as well as for the Icelandic economy, is gradually growing along with increased tourism. During recent decades, however, the Icelandic wilderness has undergone rapid change, greatly affecting the quality of the wilderness (i.e. Ólafsdóttir and Runnström 2011a, b; Taylor 2011). In order to predict how such changes will affect the future of Icelandic tourism and the tourism industry, quantitative as well as qualitative assessment of Icelandic wilderness resources is critical. So far, several attempts have been undertaken to map the Icelandic wilderness resources. The attempts to map existing wilderness based on the definition of designated wilderness areas according to the Icelandic Act no. 44/1999 on Nature Conservation include (i) official mapping using 5 km distance from major roads (i.e. The Environment Agency of Iceland and National Land Survey of Iceland 2009); (ii) an assessment of Iceland's designated statutory wilderness using proximity analysis (i.e. Ólafsdóttir and Runnström 2011a); and (iii) mapping areas outside visibility of anthropogenic structures, using a viewshed analysis (i.e. Ólafsdóttir and Runnström 2011a, b). In the present study an attempt has been made to map perceived wilderness areas by analysing areas that tourists perceive as wilderness, based on questionnaires and interviews with tourists in several tourist destinations in the southern Icelandic highlands. Efforts have been made to combine the aspects of different tourists with regard to their experience of wilderness by mapping the perception of tourists according to the purism scale and comparing their wilderness perception to the physical features in the area of study, as well as to the extent of the designated wilderness. The results of the mapping of wilderness perception reveal a major difference in opinion between the purism groups with regards to where perceived wilderness exists. Non-purists perceive almost the whole study area as wilderness, while strong purists make notably higher demands of wilderness than the official definition of designated wilderness does. This finding is supported by Higham et al. (2001) and Flanagan and Anderson (2008), indicating similar differences between different purism groups. One noteworthy difference, however, is that all structures related to power plants seem to disrupt the experience of wilderness by all purism groups visiting the Icelandic wilderness, including the non-purism group. This might be due to the barrenness of the Icelandic landscape, making anthropogenic structures particularly striking in the landscape. Another notable difference is that mountain huts do not seem to affect the experience of wilderness in any of the purism groups in the Icelandic wilderness, not even the strong purists. This can be expected to depend on the Icelandic mountain huts still being relatively small and primitive, and being well fitted into the landscape. This situation may change, as many mountain huts have been evolving into larger service centres, in order to meet the increased demands of the rapidly growing tourism in the Icelandic highlands.

11.4.2 Management Implication

The management of the world's wilderness areas is representative of many of the conflicts and challenges faced in natural resource management today. The Icelandic wilderness, whether it is viewed as an ontological reality or as an idea, is an important resource for the Icelandic tourism industry and consequently for the Icelandic economy. A public resource like wilderness can only be protected from overuse and destruction with regulation and supervision, as stated by Hendee et al. (1990). Once a wilderness area becomes known as a tourist destination, maintaining its wilderness condition becomes increasingly difficult. In order to avoid the overuse of wilderness for tourism and by other economic sectors, ambitious planning and appropriate management are critical. This includes identifying limits of growth and further development. Without such limitations, the exploitation of wilderness is unsustainable, which is against the European Parliament declaration on wilderness areas (http://www.wildeurope.org), as well as against the Icelandic government policy on sustainable development (i.e. Ministry of the Environment 2010). Recent studies (i.e. Sæþórsdóttir 2010b, 2013) highlight increased crowding within the Icelandic highlands and an accompanying reduction in the quality of wilderness experience. Given these facts, the results of this study suggest that a redirection of the non-purist group to less pristine areas in the lowlands may preserve the highlands for the market groups who have higher demands on the quality of wilderness areas. However, in order to maintain the remaining wilderness in the Icelandic highlands, increased research on the wilderness and its quality is vital. Geodiversity and biodiversity are factors likely to play large roles with regards to the quality of the Icelandic wilderness, as well as to the subjective experience of the wildness of the Icelandic highlands.

Acknowledgements The questionnaire surveys were supported by The Icelandic Ministry of Industries and Innovation, The Icelandic Tourist Board and The Icelandic Road and Coastal Administration. In addition, we are grateful to David Folkmann Drost for English correction.

References

Benediktsson, K. (2008). Moving places? The emotional politics of nature. In J. O. Bærenholdt & B. Granås (Eds.), *Mobility and place: Enacting Northern European peripheries* (pp. 205–217). Aldershot: Ashgate.

Buhalis, D. (2000). Marketing the competitive destination of the future. *Tourism Management, 21*(1), 97–116.

Carver, S., Evans, A., & Fritz, S. (2002). Wilderness attribute mapping in the United Kingdom. *International Journal of Wilderness, 8*(1), 24–29.

Cronon, W. (1998). The trouble with wilderness; or, getting back to the wrong nature. In W. Cronon (Ed.), *Uncommon ground: Rethinking the human place in nature* (pp. 69–90). New York: Norton.

Flanagan, T. S., & Anderson, S. (2008). Mapping perceived wilderness to support protected areas management in the San Juan National Forest, Colorado. *Forest Ecology and Management, 256*(5), 1039–1048.

Fredman, P., & Emmelin, L. (2001). Wilderness purism, willingness to pay and management preferences: A study of Swedish mountain tourists. *Tourism Economics, 7*, 5–20.

Gössling, S., & Hultman, J. (2006). *Ecotourism in Scandinavia, lessons in theory and practice*. Wallingford: CABI International.

Hendee, J. C., Catton, W. R., Jr, Marlow, L. D., & Brockman, C. F. (1968). *Wilderness users in the Pacific Northwest: Their characteristics, values, and management preferences* (Research paper PNW). Portland: Pacific Northwest Forest and Range Experiment Station.

Hendee, J. C., Stankey, G. H., & Lucas, R. C. (1990). *Wilderness management* (2nd ed.). Golden: North American Press.

Higham, J. (1998). Sustaining the physical and social dimensions of wilderness tourism: The perceptual approach to wilderness management in New Zealand. *Journal of Sustainable Tourism, 6*(1), 26–51.

Higham, J., Kearsley, G. W., & Kliskey, A. D. (2001). Multiple wilderness recreation management. Sustaining wilderness values – Maximising wilderness experiences. In G. Cessford (Ed.), *The state of wilderness in New Zealand* (pp. 81–93). Wellington: Department of Conservation.

Huijbens, E. H., & Benediktsson, K. (2007). Practising highland heterotopias: Automobility in the interior of Iceland. *Mobilities, 2*(1), 143–165.

Icelandic Act no. 44/1999 on Nature Conservation, section 3.

Icelandic Act no. 60/2013 on Nature Conservation, section 5.

Ísleifsson, S. R. (1996). *Ísland, framandi land* [Iceland, a strange country]. Reykjavík: Mál og menning.

Ísleifsson, S. R. (2009). Ideas of an island in the north. *Scandia, 75*(2), 97–101.

ITB (Icelandic Tourist Board). (2012). *International visitors in Iceland: Visitor survey summer 2011*. Reykjavík: Icelandic Tourist Board.

Júlíusson, Á. D. (1995). *Farmers during epidemic times*. Unpublished Ph.D. thesis, University of Copenhagen, 597 p.

Karlsson, G. (1975). Frá þjóðveldi til konungsríkis [From Commonwealth to kingdom]. In *Saga Íslands II* [Icelandic history II] (pp. 3–54). Reykjavík: Hið íslenzka bókmenntafélag.

Karlsson, G., & Kjartansson, H. S. (1994). Plágurnar miklu á Íslandi [The great epidemics in Iceland]. *Saga, tímarit Sögufélagsins, 32*, 11–74.

Kliskey, A. D., & Kearsley, G. W. (1993). Mapping multiple perceptions of wilderness in southern New Zealand. *Applied Geography, 13*, 203–223.

Landsvirkjun. (2013). *Hydro and geothermal stations*. Retrieved May 22, 2013 from http://www.landsvirkjun.com/Company/PowerStations/

Lesslie, R. G., & Taylor, S. G. (1985). The wilderness continuum concept and its implication for Australian Wilderness Preservation Policy. *Biological Conservation, 32*, 309–333.

Lupp, G., Höchtl, F., & Wende, W. (2011). Wilderness- A designation for Central European landscapes? *Land Use Policy, 28*, 594–603.

Ministry of the Environment. (2010). *Velferð til framtíðar. Sjálfbær þróun í íslensku samfélagi. Áherslur 2010–2013* [Welfare for the future. Iceland's national strategy for sustainable development. Emphasis 2010–2013]. Reykjavík: Ministry of the Environment.

Ministry of the Environment & The National Planning Agency. (1999). *Miðhálendi Íslands, Svæðisskipulag 2015* [The Central Highlands, regional plan 2015]. Reykjavík: Ministry of the Environment/The National Planning Agency.

Miriam, R. (2006). *Iceland's ecological crisis: Large scale renewable energy and wilderness destruction*. Retrieved March 24, 2013 from http://www.ru.org/ecology-and-environment/iceland-s-ecological-crisis-large-scale-renewable-energy-and-wilderness-destruction.html

Mohsin, A. (2005). Tourist attitudes and destination marketing – The case of Australia's Northern Territory and Malaysia. *Tourism Management, 26*(5), 723–732.

Ólafsdóttir, R., & Runnström, M. C. (2011a). How wild is Iceland? Assessing wilderness quality with respect to nature based tourism. *Tourism Geographies, 13*(2), 280–298.

Ólafsdóttir, R., & Runnström, M. C. (2011b). Endalaus víðátta? Mat og kortlagning íslenskra víðerna [Endless expanses, Assessing and Mapping Icelandic Wilderness]. *Náttúrufræðingurinn, 81*(2), 57–64.

Ólafsdóttir, R., & Runnström, M. C. (2013). Assessing hiking trails condition in two popular tourist destinations in the Icelandic highlands. *Journal of Outdoor Recreation and Tourism, 3–4*, 57–67.

Oslund, K. (2011). *Iceland imagined. Nature, culture, and storytelling in the North Atlantic*. London: University of Washington Press.

Pálsdóttir, S. (2005). Landsvirkjun : fyrirtækið, framkvæmdir þess og hlutverk [Landsvirkjun: The company, practices and function]. In S. Pálsdóttir (Ed.), *Landsvirkjun 1965–2005: Fyrirtækið og umhverfi þess* [Landsvirkjun 1965–2005: The company and its environment] (pp. 13–110). Reykjavík: Hið íslenska bókmenntafélag.

Schreyer, R. (1976). Sociological and political factors in carrying capacity decision-making. In *Proceedings, third resources management conference, April 29–30, U.S. Department of the Interior, National Park Service, Southwest Region, Fort Worth/Santa Fe, 1976* (pp. 228–258).

Stankey, G. H. (1973). *Visitor perception of wilderness recreation carrying capacity* (Vol. USDA Forest Service Research Paper INT – 142.) Ogden: Intermountain Forest & Range Experiment Station, Forest Service, U.S. Dept. of Agriculture, 61 p.

Stankey, G. H., & Schreyer, R. (1987). *Attitudes toward wilderness and factors affecting visitor behaviour: A state-of-knowledge review*. General Technical Report, Intermountain Research Station, USDA Forest Service, No. INT-220, pp. 246–293.

Statistics Iceland. (2013a). *Ferðamál, samgöngur og upplýsingatækni. Ferðaþjónusta. Útflutningur vöru og þjónustu eftir atvinnugreinum* [Tourism, transportation and information technology. Tourism industry. Export of goods and services by industries]. Retrieved June 26, 2013 from http://www.hagstofa.is/Hagtolur/Ferdamal-samgongur-UT/Ferdathjonusta

Statistics Iceland. (2013b). *Mannfjöldi* [Population]. Retrieved March 24, 2013 from http://www.hagstofa.is/Hagtolur/Mannfjoldi/

Sæþórsdóttir, A. D. (2008). Wilderness tourism in Iceland: Threats and opportunities. In C. M. Hall, D. K. Müller, & J. Saarinen (Eds.), *Nordic tourism: Issues and cases*. Bristol: Channel View Publications.

Sæþórsdóttir, A. D. (2010a). Planning nature tourism in Iceland based on tourist attitudes. *Tourism Geographies, 12*(1), 25–52.

Sæþórsdóttir, A. D. (2010b). Tourism struggling as the wilderness is developed. *Scandinavian Journal of Hospitality and Tourism, 10*(3), 334–357.

Sæþórsdóttir, A. D. (2011). *Wilderness tourism in Iceland – land use and conflicts with power production*. PhD thesis, The Geographical Society of Northern Finland and The Department of Geography, University of Oulu, Oulu.

Sæþórsdóttir, A. D. (2012a). Tourism and power plant development: An attempt to solve land use conflicts. *Tourism Planning & Development, 9*(4), 339–353.

Sæþórsdóttir, A. D. (2012b). *Ferðamennska á miðhálendi Íslands: Staða og spá um framtíðarhorfur* [Tourism in the central Icelandic highlands. Current situation and future expectations]. Reykjavík: Faculty of Life and Environmental Sciences, University of Iceland.

Sæþórsdóttir, A. D. (2013). Managing popularity: Changes in tourist attitudes to a wilderness destination. *Tourism Management Perspectives, 7*, 47–58.

Sæþórsdóttir, A. D., & Stefánsson, Þ. (2009). *Töluleg greining á viðhorfum ferðamanna á Kili* [Statistical analyse on tourists' attitudes in Kjölur]. Reykjavík: Institute of Geography and Tourism, University of Iceland.

Sæþórsdóttir, A. D., Hall, C. M., & Saarinen, J. (2011). Making wilderness: Tourism and the history of the wilderness idea in Iceland. *Polar Geography, 34*(4), 249–273.

Taylor, V. F. (2011). *GIS assessment of Icelandic wilderness from 1936–2010*. Unpublished MS thesis, University of Iceland. Retrieved March 24, 2013 from http://skemman.is/item/view/1946/9876

The Environment Agency of Iceland & National Land Survey of Iceland. (2009). *A map of Icelandic wilderness*.

The Wilderness Act. (1964). Section 2. (c) Definition of wilderness. Public Law 88–577 (16 U.S. C. 1131–1136). 88th Congress, Second Session, September 3, 1964.

Thórhallsdóttir, T. E. (2007). Environment and energy in Iceland: A comparative analysis of values and impacts. *Environmental Impact Assessment Review, 27*(6), 522–544.

Tuan, Y. F. (1990). *Topophilia: A study of environmental perception, attitudes, and values*. New York: Columbia University Press.

Van den Berg, A. E., & Koole, S. L. (2006). New wilderness in the Netherlands: An investigation of visual preferences for nature development landscapes. *Landscape and Urban Planning, 78*, 362–372.

Vistad, O. I. (1995). *I skogen og I skolten. Ein analyse av friluftsliv, miljöoppleving, påverknad og forvaltning i Fremundsmarka, med jamföringar til Rogen og Långfjället* [In woods and mountains. An analysis of outdoor recreation, environmental experience, impact and management in Fremundsmarka, compared with Rogen and Långfjället]. Trondheim: Norwegian University of Science and Technology.

Wallsten, P. (1988). *Rekreation i Rogen. Tillämpning av en planeringsmetod for friluftliv*. Kommitrapport 1988:2. Kommiteen for miljövern, Universitetet iTrondheim.

Williams, D. R. (2002). The social construction of arctic wilderness: Place meanings, value pluralism, and globalization. In A. E. Watson, L. Alessa, & J. Sproull (Eds.), *USDA Forest Service proceedings RMRS-P-26*. Wilderness in the Circumpolar North: Searching for compatibility in ecological, traditional, and ecotourism values, Anchorage, Alaska, May 15–16, 2001, pp. 120–132.

Chapter 12
Is There Something Wild in Austria?

Christoph Plutzar, Karin Enzenhofer, Flora Hoser, Michael Zika, and Bernhard Kohler

Abstract This chapter presents the first spatially explicit wilderness map for the Austrian territory. This is modelled using the spatial patterns of four aspects of wilderness, an approach developed by the Australian Heritage Commission: remoteness from settlements, remoteness from access, apparent naturalness and biophysical naturalness. In order to combine these four layers we applied two approaches, which reflect two different aspects of wilderness quality, namely a weighted overlay and a minimum operator. These two approaches were merged to gain a spatially explicit estimation of the wilderness continuum for all of Austria. By applying two different thresholds to the continuum, we identified core as well as extended areas, which can be considered as wild areas with high potential for wilderness. In total 1.98 % and 6.16 % of the country can be classified as core and extended areas, respectively. The vast majority of these areas are located in mountain regions with higher elevations occurring especially in the western parts of Austria. Despite some shortcomings of this approach, e.g. the lack of data describing extensive land use like grazing, we hope that this assessment can serve as a policy- and management relevant tool to improve wilderness quality in Austria.

Keywords Wilderness mapping • Wilderness continuum • Austria

12.1 Introduction

Due to their long history of human colonization, Central European landscapes hold only a few remaining areas of true wilderness. The shift in nature conservation that has occurred over recent decades, changing from a species-focused point of view to a more ecosystem-oriented approach, has drawn attention to the importance and value of wilderness areas, intact ecosystems and full functioning of ecological

C. Plutzar (✉)
Faculty of Interdisciplinary Studies, Institute of Social Ecology, Alpen Adria University, Vienna, Austria
e-mail: christoph.plutzar@aau.at

K. Enzenhofer • F. Hoser • M. Zika • B. Kohler
WWF Austria, Vienna, Austria

S.J. Carver, S. Fritz (eds.), *Mapping Wilderness*,
DOI 10.1007/978-94-017-7399-7_12

processes. While only 1 % of the total land territory of Europe is currently protected as wilderness, numerous pristine or near-natural areas that should be protected as Europe's Natural Heritage are suffering from increasing intensification and land-use changes.

Beyond protecting existing wilderness, there is a high potential for 'rewilding' the landscape and restoring the ecological processes in Europe through upcoming land-use changes and demographic transitions (Secretariat of the Convention on Biological Diversity 2010). These 'rewilding areas' or secondary wilderness are a unique conservation opportunity to establish new wilderness for future generations. This goal entails that human influence be pushed back and a 'non-intervention' management be established. In addition, wildlife reintroduction programmes should – where possible – be considered to allow natural processes to determine the composition of native habitats and species.

This awareness led to a resolution of the European Parliament in 2009 to improve protection and funding for wilderness in Europe. In the same year, a conference on Wilderness and Large Natural Habitat Areas was organized through the Wild Europe Initiative, an initiative on wilderness incorporating European environmental NGOs and European Commission.

Following the conference, the Austrian Ministry of Environment placed the idea of wilderness at the heart of its new National Park strategy (endorsed in 2010), declaring that all Austrian national parks shall henceforth focus on ecological process management in their core zones explicitly referred to as "wilderness". Austria is located at the centre of Europe and the majority of its area (83,900 km^2) is dominated by the Alps. These mountain ranges still offer many aspects connected to wilderness, showing a considerable amount of wild areas. Wild areas are known to keep many facets of wilderness (Wild Europe Initiative 2012), hence these areas have both an intrinsic value and moreover a high potential to become – by changing current land use – secondary wilderness regions (Kohler et al. 2012). Nonetheless, there exists only one official wilderness area in Austria, the "Wilderness Dürrenstein", approved by the IUCN in 2003 with a total size of approx. 3500 ha, including 400 ha of untouched forest.

Thus, there is a substantial need to identify existing regions of high wilderness value as well as areas suited for wilderness in scenarios assuming policy relevant to lowering human impact. These areas could serve to establish more protected areas meeting the wilderness criteria of IUCN 1b (Dudley 2008) and also of Wild Europe (Wild Europe Initiative 2012). The latter is aimed specifically at the small-structured land-use situation on the densely populated continent of Europe.

To meet the demand for spatially explicit information, we are now able to present a GIS-based assessment of Austria's wilderness quality, based on the wilderness continuum concept, a concept initially developed by Roderick Nash in the 1960s (Nash 2001) and implemented by Lesslie and Taylor (1985) in the Australian National Wilderness Inventory. Based on this idea, various methods were applied to assign each locality of a study area a quantitative wilderness quality index score, indicating and distinguishing relative wildness on a continuous scale. European cases for this approach have been applied to several regions, for example, The

United Kingdom (Carver et al. 2002), Scotland (Carver et al. 2012), the Alps (Kaissl 2002) and even the entirety of European territory (Fisher et al. 2010; Kuiters et al. 2013). These examples have proven the feasibility and utility of wilderness continuum mapping.

12.2 Materials and Methods

In order to model the wilderness continuum, we used the approach of Lesslie et al. (1988), which distinguishes four different aspects of wilderness: (1) remoteness from settlement (remoteness from places of permanent habitation); (2) remoteness from access (remoteness from constructed vehicular access routes like roads and railways; (3) apparent naturalness (the degree to which the landscape is free from the presence of the permanent structures of modern technological society); and (4) biophysical naturalness (the degree to which the natural environment is free from biophysical disturbance caused by the influence of modern technological society) (See Chap. 2).

Similar to Fritz et al. (2000) we estimated and combined these four indicators using a multi-criteria evaluation (MCE) framework implemented in a Geographic Information System (GIS). We used ArcGIS 10.0 (ESRI 2011) and its ModelBuilder-tool to calculate weighted distance decay models with a spatial resolution of 100 m using the following input data sets.

12.2.1 Remoteness from Settlement

A map of soil sealing (Kopecky and Kahabka 2009) served as a proxy for settlements. This layer indicates the percentage of sealed area per grid cell and was derived using satellite images and remote sensing techniques. Areas without information due to cloud cover were filled using CORINE land cover (Coordinated Information on the European Environment, EEA-ETC/LUSI 2007).

To assess the 'remoteness from settlements', a weighted Path Distance to places indicating sealed soil was calculated. The Path Distance was favoured over the Euclidian distance because it considers topographical surface conditions. As the first step, the Path Distance was calculated using a Digital Elevation Model (Jarvis et al. 2008) as surface grid. To obtain weights, the grid layer 'sealed soil' was converted to points and a point kernel density was calculated. In the next step, a weighted sum was used to overlay the Path Distance and the kernel weights to gain the weighted distances. In the final step, we performed a linear stretch to receive values between 0 and 1 (0: lowest wilderness quality, 1: highest wilderness quality).

12.2.2 *Remoteness from Access*

We used data from the Open Street Map (OSM, Geofabrik 2012) as input to calculate traffic-weighted Path Distance models. The road lines served as street layer (proxy for private transport), while sections in tunnels were excluded, supposing that adjacent areas cannot be accessed by persons in the tunnels. Transport points indicated stops (proxy for public transport). We assigned weights to each class of these layers (Table 12.1), higher weights indicating a higher negative effect on the wilderness quality, and calculated weighted Path Distances for each weight separately. We then overlaid these layers using a weighted overlay as well as a minimum operator (for further explanation see Sect. 12.2.3) differentiating between private and public transport. In the next step, these two layers were combined and stretched between 0 and 1, yielding the final result for the aspect of remoteness from access.

Table 12.1 Weights for the several input layers used for remoteness from access and apparent naturalness

Class	Weight
Roads	
Bridleway	1
Cycleway	1
Footway	1
Living street	3
Motorway	5
Motorway link	3
Path	1
Pedestrian	2
Primary	5
Primary link	3
Residential	3
Road	4
Secondary	4
Secondary link	3
Service	2
Steps	1
Tertiary	3
Track	2
Track grade1	2
Track grade2	2
Track grade3	1
Track grade4	1
Track grade5	1
Trunk	5
Trunk link	3
Unclassified	1
Unknown	1

(continued)

Table 12.1 (continued)

Class	Weight
Transport points	
Aerialway station	2
Airfield	3
Airport	5
Bus station	1
Bus stop	1
Ferry terminal	2
Helipad	2
Railway halt	3
Railway station	4
Taxi rank	1
Tram stop	1
Power lines	
Cable	
Line	2
Minor cable	
Minor line	1
Points of interest	
Alpine hut	3
Restaurant	3
Ruins	3
Shelter	1
Tower	3
Railways	
Cable car	3
Chair lift	3
Drag lift	2
Funicular	3
Light rail	2
Miniature railway	2
Monorail	2
Narrow gauge	2
Rail	3
Subway	
Tram	1
Buildings	3
Skiing areas	4
Power stations	
Pole	1
Station	5
Station fossil	5
Station nuclear	5

(continued)

Table 12.1 (continued)

Class	Weight
Station solar	4
Station water	
Station wind	4
Substation	3
Hydroelectric power stations	
Storage power plants	5
Transverse structures	1
Run-of-river plants:	
standard operating capacity (GWh)	Weight
<486000	1
<850000	2
<1221600	3
<1617400	4
<1967600	5

12.2.3 Apparent Naturalness

Similar to the remoteness from access, weighted distance-decay functions were calculated using data on human infrastructure and artefacts as inputs: skiing areas (Umweltbundesamt 2012), hydroelectric power stations (Walder and Litschauer 2010), other power stations, power lines, alpine huts & shelters, the railway network and buildings (all Geofabrik 2012). We assigned weights to the different classes of the input layers (Table 12.1), calculated Path Distances followed by a weighted sum or a minimum operator (see Sect. 12.3) and a final linear stretch.

12.2.4 Biophysical Naturalness

This aspect of the wilderness quality index reflects the degree to which an area is free from the biophysical disturbances of human society. Various factors can be included here, e.g. land-use relevant activities (such as farming, forestry, fertilization and grazing) or even remote influences like emissions.

Due to a lack of adequate land use data, we used the CORINE land cover data set (EEA-ETC/LUSI 2007) as proxy, applying weights according to the degree of naturalness of land cover (Table 12.2). CORINE is a product of the European Environment Agency covering all of the EU27 territory and offering a standardized and hierarchical classification system. In wooded areas the degree of hemeroby (Grabherr et al. 1998) was included additionally. The concept of hemeroby measures the degree of naturalness of ecosystems and is used in ecological sciences.

Table 12.2 Weights for the CORINE land cover data set

Land cover class	Weight
Continuous urban fabric	5
Discontinuous urban fabric	4
Industrial or commercial units	5
Road and rail networks and associated land	4
Port areas	4
Airports	5
Mineral extraction sites	4
Dump sites	5
Construction sites	4
Green urban areas	3
Sport and leisure facilities	3
Non-irrigated arable land	3
Vineyards	3
Pastures	3
Annual crops associated with permanent crops	3
Complex cultivation patterns	3
Land principally occupied by agriculture, with significant areas of natural vegetation	3
Pastures	3
Broad-leaved forest	1–3[a]
Coniferous forest	1–3[a]
Mixed forest	1–3[a]
Natural grasslands	1
Moors and heathland	1
Transitional woodland-shrub	1
Bare rocks	1
Sparsely vegetated areas	1
Glaciers and perpetual snow	1
Inland marshes	1
Peat bogs	1
Water courses	1
Water bodies	1

[a] Weights using degree of hemeroby

12.3 Integration of Intermediate Results

For the integration of all intermediate results described above (remoteness from settlement, remoteness from access, apparent naturalness and biophysical naturalness), we followed two distinct approaches. To obtain an overall estimation of wilderness quality, we used a weighted overlay, similar to Carver et al. (2012). This approach considers all factors within a certain radius of a given location, calculating the weighted average. For finding weights (Table 12.3), we drew on Carver et al.

Table 12.3 Weights of the four intermediate layers

Layer	Weight
Remoteness from settlement	4
Remoteness from access	3
Apparent naturalness	3
Biophysical naturalness	1

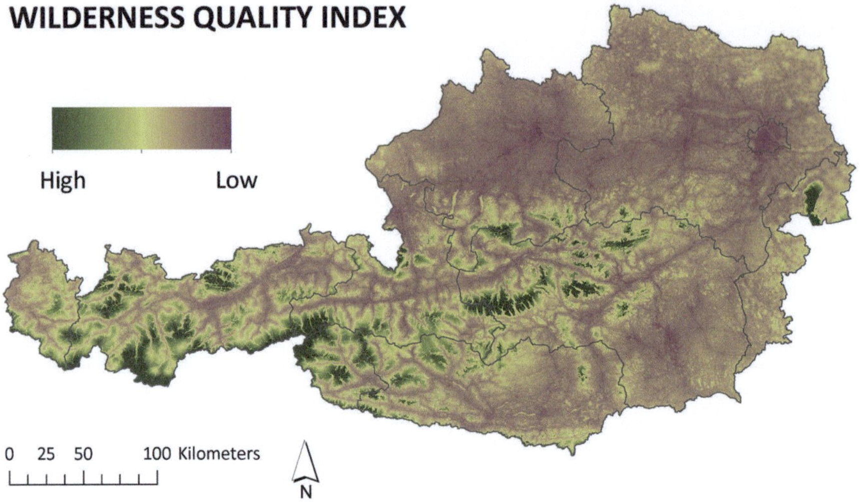

Fig. 12.1 Austrian wilderness continuum, combining a weighted overlay approach and a minimum operator approach

(2012), but had to adapt the figures for this study. We assigned the weights using a best guess considering spatial and thematic accuracy, underlying richness of information and local relevance.

This method is suited for highly populated areas, such as most European landscapes, and differs from the Australian approach (Lesslie et al. 1988; Lesslie and Maslen 1995), which only takes the most important factor into account (Fritz et al. 2000). In the case of Austria, this method tends to underestimate the influence of single facilities in remote areas (like alpine huts), because they accumulate much less weight compared to crowded localities. To be able to consider such facilities in these sensitive areas, we adapted the Australian approach and applied a so-called 'minimum operator' (which corresponds to a logical 'and'). As a consequence, for each locality the smallest and hence most influential distance value was taken into account. This minimum operator was applied using the first three aspects of the wilderness quality. The result thus reached was divided by the biophysical naturalness stretched between 0 and 1.

To obtain a final spatially explicit estimation of the wilderness quality index for all of Austria, we calculated the average of these two layers (Fig. 12.1).

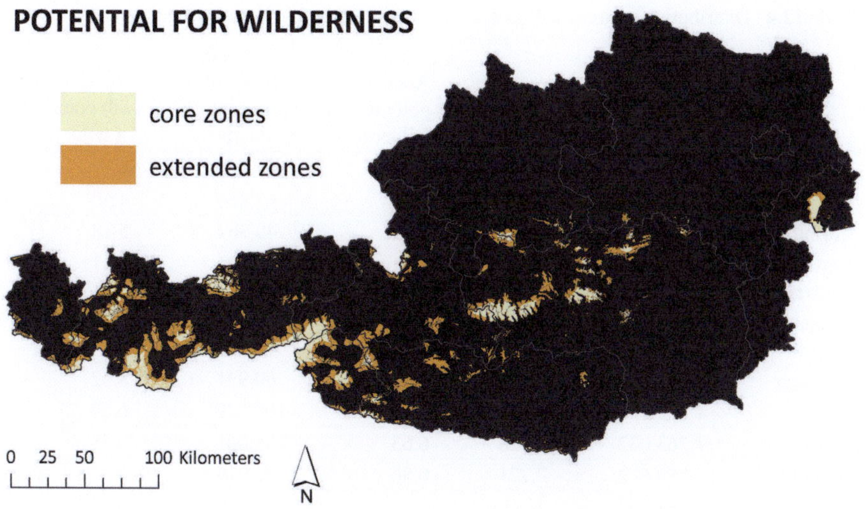

Fig. 12.2 Core and extended zones showing potential for wilderness in Austria

12.4 Delineation of Areas with Potential for Wilderness

The wilderness continuum provides valuable information with which to evaluate potential wilderness areas. However, a delineation of areas with a high potential for wilderness is also desirable for special tasks. To achieve this goal, two (arbitrary) thresholds were applied to designate such areas. To identify core areas, a threshold of 0.52 was determined, whereas a threshold of 0.39 was determined for extended areas. We sampled 100 wilderness quality index values in 500 m distance to 10 randomly chosen single objects in the alpine regions (alpine huts) and calculated the mean to derive the threshold for core areas. We proceeded analogously using 100 sample points in a distance of 2500 m to skiing areas to receive the threshold for extended areas.

The result shows that 1.98 % of Austrian territory can be considered as core areas for wilderness potential and 6.16 % show extended potential for wilderness (Fig. 12.2, Table 12.4). It must be noted that Lake Neusiedl in the east of Austria represents a considerable portion of these areas, since we did not exclude water bodies from this study.

12.5 Land Use Change Scenario

The approach presented here reflects the potential for wilderness under recent land use. To estimate the potential under changed land use, a discontinuation of certain land use activities was simulated. The simulation excluded relevant input data

Table 12.4 Distribution of areas with potential for wilderness by Austrian federal states

Federal state	Total area	Areas with potential for wilderness	Areas with potential for wilderness	Extended areas with potential for wilderness	Extended areas with potential for wilderness
	ha	*ha*	%	*ha*	%
Burgenland	395,877	11,632	2.94	17,834	4.50
Carinthia	953,513	6,431	0.67	49,369	5.18
Lower Austria	1,917,837	170	0.01	2,141	0.11
Upper Austria	1,197,522	1,908	0.16	9,723	0.81
Salzburg	715,378	22,005	3.08	60,931	8.52
Styria	1,639,656	37,854	2.31	101,729	6.20
Tyrol	1,263,032	83,728	6.63	258,061	20.43
Vorarlberg	259,672	2,221	0.86	16,915	6.51
Vienna	41,463	0	0	0	0
AUSTRIA	8,383,954	165,952	1.98	516,706	6.16

Table 12.5 Distribution of areas with potential for wilderness by Austrian federal states using the land use change scenario

Federal state	Total area	Areas with potential for wilderness	Areas with potential for wilderness	Extended areas with potential for wilderness	Extended areas with potential for wilderness
	ha	*ha*	%	*ha*	%
Burgenland	395,877	11,713.65	2.96	18,008	4.55
Carinthia	953,513	14,980.58	1.57	70,964	7.44
Lower Austria	1,917,837	175.41	0.01	4,942	0.26
Upper Austria	1,197,522	2,038.20	0.17	14,241	1.19
Salzburg	715,378	25,381.29	3.55	75,382	10.54
Styria	1,639,656	50,942.28	3.11	141,618	8.64
Tyrol	1,263,032	152,821	12.10	367,301	29.08
Vorarlberg	259,672	1,443	0.56	29,513	11.37
Vienna	41,463	0	0	0	0
AUSTRIA	8,383,954	259,496	3.10	721,973	8.61

(skiing areas, forest tracks, alpine huts and cable cars) from the model, and the whole model was rerun. Some of the alpine huts were associated with 'aerial rope-way stations' and 'helipads', so these facilities had to be excluded as well. Additional areas with potential for wilderness were added to the existing set of areas.

The spatial patterns of the results show marginal changes, there is an increase of core areas with potential for wilderness to 3.10 % and an increase to 8.61 % for extended areas with potential for wilderness (Table 12.5). Figure 12.3 shows the difference for the core zones.

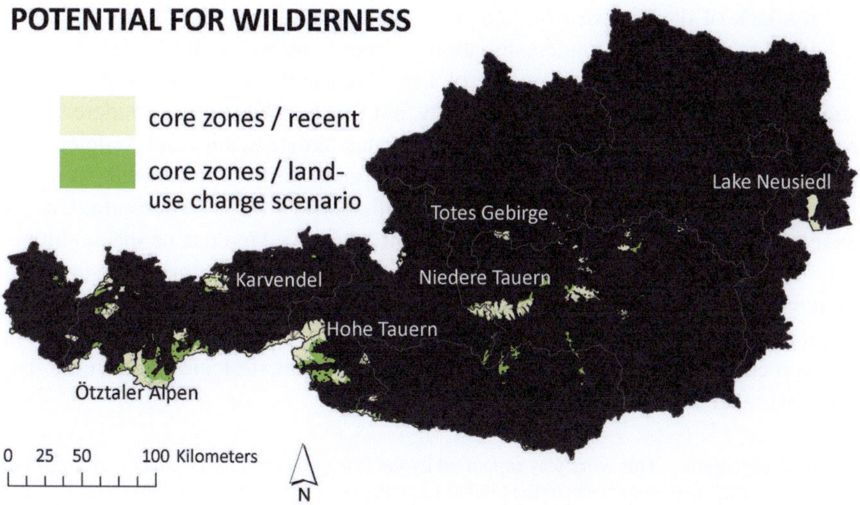

Fig. 12.3 Core zones of potential for wilderness in Austria under recent conditions and under a land use change scenario, assuming the closure of skiing areas, forest tracks, alpine huts and cable cars

12.6 Conclusion

The spatial pattern of the Austrian wilderness continuum shows that mountain ranges are favoured over lowlands. Areas with high wilderness quality are located especially in the western parts of Austria, for example the mountain regions Hohe Tauern, Niedere Tauern, Ötztaler Alpen, Lechtaler Alpen, Karwendel and Totes Gebirge. One exception is the large body of water of Lake Neusiedl, situated in the east at the border to Hungary. As expected, the populated regions of Vienna, Lower Austria, Upper Austria, the south-western parts of Styria and the large alpine valleys show consistently low wilderness quality values. This result was to be expected inasmuch as in Central European landscapes usually land-use intensity as well as most human activities decline with increasing altitudes.

Nevertheless, we are able to present this effect on a quantitative basis, corroborating the importance of alpine habitats for preserving natural processes and services on a large scale. Moreover, this approach is able to provide a point of departure for comparing the level of naturalness of different regions and localities, considering various aspects of anthropogenic disturbances. Detailed local studies could offer scenarios for how to protect existing aspects of wilderness as well as for how to change recent management and land use to develop wilderness in a sustainable way. Although the land use change scenario result shows only a small increase in the total amount of potential for wilderness (1.98–3.10 % for core zones and 6.16–8.61 % for extended areas respectively), some areas like Ötztaler Alpen and Hohe Tauern would see a considerable growth of wild land. It should be noted that the high wilderness quality value of Lake Neusiedl is a consequence of the input data used. We faced a lack of data focused on human activities on lakes – like ferries, sailing or fishery – resulting in an underestimation of human impact in freshwater habitats. Because of the national importance of Lake Neusiedl and its National Park, we decided not to exclude lakes for this study, but this bias has to be considered when reviewing the result and highlights the importance of data quality and completeness. It is clear that the assessment given here is missing several factors that would be important for a full and extensive evaluation of Austria's wilderness continuum. For example, grazing or hunting, which represent extensive land use or special human activities both affecting wilderness quality, are reported on administrative units and therefore lacking sufficient spatially explicit data sets.

Nonetheless, we hope that our work can serve as a spatially explicit tool to help develop conservation policy- and management-relevant strategies that, in the long run, will make Austria a wilder place.

Acknowledgements This work was supported by the European Research Council Starting Grant 263522 (LUISE) and contributes to the Global Land Project (www.globallandproject.org).

References

Carver, S., Evans, A., & Fritz, S. (2002). Wilderness attribute mapping in the United Kingdom. *International Journal of Wilderness, 8*, 24–29.

Carver, S., Comber, A., McMorran, R., & Nutter, S. (2012). A GIS model for mapping spatial patterns and distribution of wild land in Scotland. *Landscape and Urban Planning, 104*, 395–409.

Dudley, N. (Ed.). (2008). *Guidelines for applying protected area management categories*. Gland: IUCN.

EEA-ETC/LUSI. (2007). *CLC2006 technical guidelines* (EEA technical report 17). Luxembourg: Office for Official Publications of the European Communities. Available at: http://www.eea.europa.eu/publications/technical_report_2007_17. Accessed 11 July 12.

ESRI. (2011). *ArcGIS desktop: Release 10*. Redlands: Environmental Systems Research Institute.

Fisher, M., Carver, S. Kun, Z., McMorran, R., Arrell, K., & Mitchell, G. (2010). *Review of status and conservation of wild land in Europe*. Project commissioned by the Scottish Government.

Fritz, S., Carver, S., & See, L. (2000). New GIS approaches to wild land mapping in Europe. In S. F. McCool, D. N. Cole, W. T. Borrie, & J. O'Loughlin (Eds.), *Wilderness science in a time of change conference – Volume 2: Wilderness within the context of larger systems* (pp. 120–127), Missoula.

Geofabrik. (2012). *OpenStreetMap-Shapefiles*. Available at http://www.geofabrik.de/data/shapefiles.html. Accessed 18 Sept 2012.

Grabherr, G., Koch, G. Kirchmeir. H., & Reiter, K. (1998). *Hemerobie österreichischer Waldökosysteme*. Innsbruck: Österreichische Akademie der Wissenschaften, Veröffentlichungen des österreichischen MaB-Programms.

Jarvis, A., Reuter, H. I., Nelson, A., & Guevara, E. (2008). *Holefilled seamless SRTM data V4*. International Centre for Tropical Agriculture (CIAT). Available at http://srtm.csi.cgiar.org. Accessed 2 Aug 2012.

Kaissl, T. (2002). *Mapping the wilderness of the Alps – A GIS-based approach*. Diploma thesis at the University of Vienna, Vienna.

Kohler, B., Laßnig, C., & Zika, M. (2012). *Wildnis in Österreich? Herausforderungen für Gesellschaft, Naturschutz und Naturraummanagement in Zeiten des Klimawandels*. Purkersdorf: Österreichische Bundesforste AG (ÖBf AG), Kompetenzfeld Naturschutz.

Kopecky, M., & Kahabka, H. (2009). *Updated delivery report European mosaic*. Available at http://www.eea.europa.eu/data-and-maps/data/eea-fast-track-service-precursor-on-land-monitoring-degree-of-soil-sealing-100m-1. Accessed 14 Sept 2012.

Kuiters, A. T., van Eupen, M., Carver, S., Fisher, M., Kun, Z., & Vancura, V. (2013). *Wilderness register and indicator for Europe final report* (p. 92). EEA Contract No: 07.0307/2011/610387/SER/B.3.

Lesslie, R., & Maslen, M. (1995). *National wilderness inventory: Handbook of principles, procedures and usage* (2nd ed.). Canberra: Australian Heritage Commission.

Lesslie, R. G., & Taylor, S. G. (1985). The wilderness continuum concept and its implications for Australian wilderness preservation policy. *Biological Conservation, 32*(4), 309–333.

Lesslie, R. G., Mackey, B. G., & Preece, K. M. (1988). A computer-based method for the evaluation of wilderness. *Environmental Conservation, 15*, 225–232.

Nash, R. (2001). *Wilderness and the American mind* (4th ed.). New Haven/London: Yale University Press.

Secretariat of the Convention on Biological Diversity. (2010). *Global biodiversity outlook 3*. Montréal: Secretariat of the Convention on Biological Diversity.

Umweltbundesamt. (2012). *Skigebiete 2008*. Available at http://www.umweltbundesamt.at/umweltsituation/umweltinfo/opendata/oed_naturschutz/. Accessed 14 July 2012.

Walder, C., & Litschauer, C. (2010). *Ökomasterplan Stufe II – Schutz für Österreichs Flussjuwele! Zustand und Schutzwürdigkeit der österreichischen Fließgewässer mit einem Einzugsgebiet größer 10 km*^{2}. Wien: WWF Österreich.

Wild Europe Initiative. (2012). *A working definition of European wilderness and wild areas*. Available at http://www.panparks.org/sites/default/files/docs/iyw/Definition-of-wilderness-for-Europe.pdf. Accessed: 16 Feb 2013.

Chapter 13
Concluding Remarks

Stephen J. Carver and Steffen Fritz

Abstract Global estimates of wilderness reserves have declined rapidly over the last 300 years, largely due to human population growth and demand for food and resources. And human population, now at seven billion people looks set to rise further still to nine billion or more by 2100. Sustainability and resilience are not just words that apply to resources, but to the whole planet including wilderness. Our concluding chapter makes cogent and clear arguments in favour of wilderness protection and rewilding as one means of maintaining a healthy global ecosystem, contrary to some of the ideas of the green modernist movement that puts people at the centre and which believes that technology and human ingenuity will come to the rescue. While some of the policy mechanisms to do this are already in place (e.g. REDD+) we suggest that these require better and stricter application informed by mapping campaigns. Despite wilderness being a largely fuzzy concept, lines on maps are needed to protect natural ecosystems and their wildlife. We conclude with a review of the chapters in the book and how they, together, address these concerns, before making some predications as to future developments.

Keywords Sustainability and wilderness • Futures

> In wildness is the preservation of the world (Henry David Thoreau, 1862)

It is over 150 years since Thoreau penned these words, yet they still have strong resonance today. In the intervening century and a half we have seen the human population increase almost 6 fold from somewhere just over 1.2 billion to 7 billion in 2012, with most of us living in urban areas. Over roughly the same time period, the theories of Thomas Malthus on population checks from famine and disease have largely been avoided as human ingenuity has enabled food production and distribution – regional problems notwithstanding – to keep pace with rapid population

---

S.J. Carver (✉)
School of Geography, University of Leeds, Leeds, UK
e-mail: s.j.carver@leeds.ac.uk

S. Fritz
Ecosystem Services and Management, International Institute of Applied System Analysis, Laxenburg, Austria

S.J. Carver, S. Fritz (eds.), *Mapping Wilderness*,
DOI 10.1007/978-94-017-7399-7_13

growth. Huge cities supported by industrial-scale agriculture and fisheries together with geo-resource exploitation on a hitherto unheard of scale work together to support a massively complex global human ecosystem. This is a trend that seems set to continue with a global population of 9 billion predicted by 2050; though our ability to feed this number of people without further drastically impacting natural ecosystems is by no means as certain.

Not only will there be more people to support but demand will grow for better living standards, including a more meat-based diet, better housing, consumer products, higher mobility, etc., which will place additional strain on basic key resources. This will doubtless mainly manifest itself in demand for land for living space and agriculture. Land will also be under pressure from resource extraction (oil, gas, coal, minerals, timber, water) and energy production which is set to come increasingly from renewable energy and biofuels as we approach Peak Oil.[1] Much of this demand will be driven by the burgeoning populations of developing and emerging countries in Asia, the Indian sub-continent, Africa, Central and South America together with political efforts to maintain and even increase current standards of living in the developed world.

13.1 Towards Global Sustainability with Wilderness

Without careful planning and policies to the contrary, it is easy to see how such an increase in demand will impact adversely on wilderness areas. It is therefore essential that, if we are to safeguard the future of wilderness, and following from Thoreau's dictum help save the planet, then new and innovative approaches to land use and resource management are required. Such innovative approaches should also take a global perspective. The world is becoming increasingly interconnected and what is done on one side of the planet might impact on another part. Models and tools which take such a global perspective include earth system models, integrated assessment models and economic land-use models, many of which focus on the biosphere (Havlik et al. 2011; Asselen and Verburg 2013). Some of these models allow us to examine the impact of a policy or consumer behaviour on biodiversity and protected areas. One example is the impact of certain pathways on the world forests such as that provided by the global forest report. Here the global land-use model shows that changes in consumption patterns, particularly among the most affluent, will be need to achieve zero net deforestation and forest degradation (ZNDD). This essentially means protecting wilderness in areas of high productivity without compromising other ecosystems or food security. Interestingly, such consumption changes are not so dramatic as to be either socially or practically implausible (Taylor 2011).

[1]Estimates of Peak Oil (maximum historical output per capita) vary as known reserves, new extraction technologies, variable demand and oil prices can mean output figures fluctuate but recent estimates suggest global output may peak in 2015.

In order to feed a growing global population a more sustainable use of land and resources is needed through agricultural intensification, more efficient use of limited resources through advanced IT and distribution systems (e.g. precision farming techniques, "just-in-time" distribution networks, etc.) resulting in better yield management and less wastage. Moreover, all models show that in order to maintain wilderness, quite radical changes in lifestyles and patterns of consumption are also required such as reducing the amount of meat in our diet. At the same time it is imperative that we slow population growth and restructure demography, while accepting that sustained economic growth might not always be possible across all sectors. These are all very hard economic, social and political choices, but it is essential that they are addressed if long-term global sustainability is to be achieved. Business as usual is no longer a valid option.

While we recognise that technological fixes to Malthusian problems of supply and demand have occurred in the past (e.g. the Green Revolution in the 1960s) we cannot rely wholly on the expectation that future technological advances will allow supply to keep pace with demand for food, energy, water and other strategic resources. Indeed, it can be seen in recent trends that such fixes can actually have short and medium term impacts on wilderness and other natural ecosystems.

Demand for energy is a good example. Exploration for and exploitation of increasingly difficult to get at reserves of oil, gas and coal are causing extensive damage to natural ecosystems on land and at sea; witness the impacts and ongoing debates over exploiting shale-gas, tar-sands and oil-shales. Even renewable energy has the potential to generate adverse impacts on the natural environment. Hydropower dams rivers and drowns valleys, wind energy generates visual impacts and impacts on bird-life, wave and tidal power disrupts tidal flows and marine ecosystems, biofuel monocultures have huge land requirements, displace food production and reduce biodiversity. Even industrial solar requires vast land reserves and all renewable energy sources are not carbon neutral when we consider the embodied carbon involved in the manufacture, construction and running costs. Energy efficiency and reduction is clearly the way to go together with distributed modes of generation, but population growth will nevertheless ensure that demand remains high.

There is a developing paradigm based around the ecology of the Anthropocene that maintains that all aspects of the planetary system have been and are modified and controlled by human action. The reasoning follows that wilderness is merely a human construct, a kind of imagined nature, and has no real relevance to the modern world where much of the planet is populated, farmed, exploited and therefore dominated by novel ecosystems. This is accompanied by a kind of 'humans first' philosophy and faith that technology will prevail.[2] While technological optimism is a good thing (and we should continue to research new more environmentally benign technologies) and recognising the fact that many ecosystems are heavily modified (some with distinct and obvious benefits for humans such as food production), we should also recognize at the same time that nature has a place itself. We need to recognise

[2] See for example the Breakthrough Institute http://thebreakthrough.org/

nature's right to exist and we should keep as much wilderness as is possible in places where nature can flourish without the influence of modern technological society. We need wilderness for a multitude of ecological, environmental, social, economic and cultural benefits (i.e. the ecosystem services model) and should not forget that wilderness and wild nature has its own intrinsic values that are not tied solely to human benefits. Paraphrasing from Aldo Leopold (1949)… "*The first rule of intelligent tinkering is to keep all the parts*" and this ought not just to refer to individual species, but whole landscape-scale ecosystems as well. The suggestion that we (and the planet) can survive without wilderness, intact ecosystems and abundant biodiversity very much risks throwing the baby out with the bathwater. Understanding the spatial and temporal patterns of supply and demand together with opportunities for re-thinking demand-side systems through efficiencies of use and waste minimisation is essential to gaining a better understanding of the big picture regarding population, resources and nature. All this ought to give wilderness and natural processes a place, providing as it does so many of the ecosystem services that we humans rely on.

Conserving wilderness on a global level is closely linked to other important global issues such as reducing the loss of biodiversity, maintaining ecosystem functioning, increasing the global protected areas network and stopping deforestation. In order to make sure nature and wilderness is protected it is essential that all organisations involved in conservation efforts work together in order to protect wilderness and the species living in it. Moreover conservation and protected area designation needs to be co-ordinated internationally. Recent work has shown that when using the land use change models described above it is only through international collaboration and co-ordination the Aichi Biodiversity Target 11 (adopted by the Convention on Biological Diversity) that the protected area network could be expanded to at least 17 % of the terrestrial world by 2020. The study furthermore demonstrates that international action is urgently needed to balance land-use and biodiversity conservation.

We need to be self-critical here as well, since much of this thinking about wilderness and biological biodiversity is mostly a luxury of the wealthy. We need to listen to the developing nations' view that we use wilderness and protected areas to continue to control them (Cholchester 1996). In order to address such criticism we need to draw a line between stopping large-scale developers from exploitation of natural resources and prohibiting poor people from the economic development the western world has already benefited from. Furthermore, we need to recognize that indigenous people and their role in protected areas deserves close scrutiny whilst being aware that 'wilderness' is a western construct and one that is often new to indigenous people.

As already emphasised in Chap. 2, the impact on wilderness is defined as the 'degree of influence from modern technological society' and not of traditional and indigenous societies. Cholchester (1996) outlines a number of lessons learned. First, it is recognized that successful conservation can only be achieved if indigenous peoples' rights to own their territory is granted, if indigenous representatives' institutions are recognized and if mechanisms are in place which ensure the involvement of marginal sectors in ways that do not undermine traditional decision making. Colchester furthermore highlights that support of government institutions which actually respect these principles is needed in order to help to protect from external pressures. He even goes as far to suggest that a separate IUCN category which encompasses such principles is needed and highlights that only very few countries have national legislation which permits rights of indigenous people within protected areas (Cholchester 1996). Nonetheless, this does not give indigenous peoples the right to exploit wilderness areas using modern technology in a manner and extent that produces large scale irreversible impacts, rather it is important to support indigenous cultures protect and value their lands and natural history.

The rich nations therefore need to promote and financially support local communities in developing countries which protect nature. One way of addressing this issue is the REDD+ mechanism and providing Payments for Ecosystem services (PES). REDD+ provides an opportunity to achieve large-scale emissions reductions and achieve conservation objectives by economically valuing the role forest ecosystems play in providing ecosystem services. The mechanism allows intact forests to compete with historically more economically viable land uses (i.e. the conversion to pasture or cropland). However, the REDD+mechanism is still under discussion by the UNFCCC it is not yet clear if the mechanism will truly result in protecting wilderness or if commercial interests are prioritized over environmental objectives in the forming of REDD+ policy.

It has to be recognized that wilderness protection is not free and it is not enough to draw a line around wild areas, but true protection in areas with human presence only works if there are economic alternatives for local communities such as eco-tourism or low impact tourism. Interestingly people like to go to the wild to see large predator species such as lions and the economic value of a national park is much higher if it contains those wild animals which only survive in a natural environment with little human impact.

13.2 Back to the Map

This brings us neatly back to the map. The spatial pattern of the world's remaining wilderness ecosystems, the landscapes and biodiversity they preserve, the ecosystem services they provide, the threats they face, is all important knowledge. This book has opened a window on the world of mapping the wild and some of the hard and often philosophical questions that go with it; to map or not to map, that is the question. Returning to our opening comments on early maps and "Here be dragons", how can there be real wilderness in the world if everywhere is mapped and therefore known? As Leopold (1949) so eloquently puts it…"*Of what avail are forty freedoms without a blank spot on the map?*"

Philosophical musings aside, once a wilderness is lost to human development, it is lost for ever. Or is, at the very best, extremely difficult to recreate depending on what exactly has been lost with it; the "Jurassic Park" model of species resurrection still being a technological day-dream. It is therefore essential that we bring to bear all the available data and technological resources to map our remaining wilderness areas in a rigorous, robust and repeatable manner such that wilderness and wild nature can be defended against those with the short-term view that only sees it as land and associated resources that can be exploited for financial gain.

It is apposite to use a personal story to illustrate this point further. Inspired by the work of Rob Lesslie, we first started experimenting with GIS and spatial data for mapping wilderness quality back in the mid-1990s. At the time we were warned by Bob Bunce[3] whom we had approached to get hold of some reliable land cover data for the UK, that we risked attracting people to the very thing we were seeking to protect, for (as eluded to above) to map is to know. Publicly identifying the wildest parts of the UK countryside might, Bob had suggested, merely result in greater visitor numbers seeking a wilderness-type experience. "*The woods are overrun and sons-of-bitches like me are half the problem*" (Fletcher 1971). We were able to rationalised such concerns – real as they may be – safe in the knowledge that we pretty much know where they are anyway and in reasoning that without detailed mapping they would be at greater risk from attrition and exploitation by human development and land use. Nevertheless, we both resisted the temptation to put a line on the map identifying the wildest areas of the country for many years thereafter because any such line would necessarily be artificial due to wilderness being a fuzzy concept dependent on definitions and individual experience and background.

More recently we have changed our mind on this, having come to realise that without that line on the map it becomes difficult to defend in front of planners, economists, politicians and policy-makers who only understand crisp decision rules based on lines on maps. No amount of warm, fuzzy, heart-felt pleas for the importance of wilderness is going to hold back the tide of development without a hard line on the map acting as a kind of battle front.

It is for this reason that various governments, agencies and NGOs around the world have started to turn to spatial information and mapping technologies to provide accurate and timely information to support their emerging policies and decision-making regarding wilderness. Numerous examples have been provided throughout this book. Global, regional and national scale maps have been produced including the global scale Human Footprint (Sanderson et al. 2002) and map of human impact on the world's oceans (Halpern et al. 2008). Continental scale mapping is available now for Europe (Kuiters et al. 2013) and there are various examples for national and regional maps available (see Henry and Husby 1994; Aplet et al. 2000; and Chaps. 10, 11 and 12 in this book). Scotland remains perhaps the world's most extensively mapped country with wilderness quality mapping available at all spatial scales from global to European, and from national to local (Fisher et al. 2010). Local park-wide maps are available for the Scottish National Parks (Carver et al. 2012) and a programme of mapping wilderness character within US National Parks has recently begun (Tricker et al. 2012; Carver et al. 2013).

[3] Then at the Institute of Terrestial Ecology, Merlewood (now at Alterra, University of Wagenigen, Netherlands)

In many cases these maps and information are being used by stakeholders to both inform decisions and defend against actions that might otherwise harm wilderness. As described in Chap. 1, the Scottish wild land maps are now part of Scottish Planning Policy (SPP) and the concept of wild land is embedded within the Third National Planning Framework (NPF3). This would not have come about if GIS-based mapping was not able to define exactly where these areas are on a map using techniques (and data) that are rigorous, robust, repeatable and defensible. This information is already being used by planning authorities and NGOs to present evidence supporting objections to proposed developments that would negatively impact on wild land areas. These include industrial wind farms in the heart of the Scottish Highlands and a gold mine in a national park. More proactively, the mapping is being used to promote wild land as a formal conservation designation and target areas for rewilding.[4]

We hope that the chapters in this book have demonstrated that quantitative mapping of what is essentially a qualitative concept, albeit of a biophysical entity, is both possible and practical. We have shown how mapping can provide the information necessary to build the foundations of a robust policy of protection (Chap. 2). We have shown that ecological connectivity and thinking on a landscape scale is required if protected area systems are to prove resilient to climate change and the onslaught on human development (Chap. 3). We have shown how data collection and new mapping techniques can help overcome some of the hitherto intractable problems in mapping wildness at meaningful levels of reliability, making our maps both robust and repeatable (Chaps. 4, 5 and 6). We have developed new ways of looking at the world that emphasise the importance of wilderness areas and allow us to visualise unseen patterns in our data (Chap. 7). We have debated the legal and philosophical aspects of mapping wilderness and why it is necessary if we are to have any semblance of a wilderness experience in the future (Chaps. 8 and 9). And finally we have shown that by mapping wilderness and wildness at different scales, using different models and data can tailor our results to suit local and regional variations in the concept and use these to inform developing policy on preserving the world's last remaining wild places (Chaps. 10, 11 and 12).

13.3 Back to the Future

So, what does the future hold? Bigger, better, faster, and in more detail? Better methods are being developed that are able to cope with and make use of better (faster) computers and better (more comprehensive, reliable and more detailed) data. Some of the chapters in this book have illustrated some of these recent and ongoing advances (see for example Chaps. 5, 6 and 7).

As we have already outlined, mapping provides a powerful tool to defend boundaries and to demarcate areas. With new technologies and mobile phones being avail-

[4] See for example the work of the John Muir Trust http://www.jmt.org/

able even to indigenous people in remote areas, we can put mapping tools into the hands of all citizens across the world to multiply the power of mapping and to give communities the necessary tools to demarcate and defend their areas from development pressure. This crowdsourcing approach to data collection, validation and verification is one way in which the future of wilderness mapping will expand and develop (see for example Chap. 6).

On the other hand wilderness mapping from space will become more affordable and accessible and internet platforms which bring these data streams together, illustrate and document, and map threads and conflicts will pay an increasingly important role (e.g. see moabi.org a platform which provides information about forest concessions in the DRC). With new remote sensing technologies we enter the era of near real-time mapping of wilderness. Only 10 or 15 years ago Google Earth seemed futuristic. We now take desktop and mobile versions of Google Maps and Google Street View for granted, and it is clear that such products are only going to increase in coverage and detail over the next few years. Maps and mapping are certainly some of the most heavily used tools on the Internet. UAVs or drones will increasingly be used to collect detailed imagery of areas of interest in areas that are difficult or dangerous to access. These will be used for monitoring change, wildlife behaviours, human activities and impacts, although issues about privacy and safety are now beginning to cause concern. Wildlife itself can become the vehicle for monitoring with increased miniaturisation of low-cost cameras and GPS units enabling the animals themselves to be the vehicle of data collection. The data collected in this fashion will be increasingly valuable for studies of wildlife movement, distribution and behaviour in wilderness areas throwing new light on to the hitherto secret lives of animals.

If we go back to the concept of Rob Lesslie on wilderness, action is needed on the entire spectrum of the scale on the wild end as well as on the urban end. Research is needed in the field of ‘leave no trace’ science wherein science becomes less and less intrusive in wilderness areas. At the same time we need to focus on places which have been modified and can be improved in terms of naturalness quality, wilderness and biodiversity. This is the burgeoning field of rewilding, a term which has come to encompass ecological restoration, re-establishing connectivity between protected areas, non-intervention management, species reintroduction and the reconnection of people with nature. More research is needed in the field of wildlife corridors and areas suitable for rewilding in particular focusing on the means and benefits of rewilding and the efficacy of ecological networks and corridors. Also urban wilderness science can play an important role in bring wildness into the city and to hearts and minds of millions. For example where are the quiet areas in the city, where is noise and light pollution highest and where are relatively the wildest areas in an urban setting? Considering that in the year 2050 there will be as many people living in urban areas as the overall world population in the year 2004 it is imperative that we maintain pockets of wild and relatively undisturbed places in the urban areas as well as in its close vicinity. More wilderness is clearly needed on all levels of the wilderness continuum, not less.

So, where do we go next? The previous few paragraphs have outlined some ongoing and recent developments in the field of wilderness mapping and associated spatial sciences. It is always tempting to reach for the crystal ball when concluding a book like this and make a few predictions, and it is, more often than not, sobering a few years later when you read back over them and realise how wrong you were! Nevertheless, we might like to think about some of the likely future developments and to this end we will leave you with the following list:

Data and tools: We have over the last 20–30 years moved from being data poor to being data rich. There is no reason to doubt why this trend should not continue right across the board from satellite imagery to qualitative data. The challenge is most likely to be how best to make good use of it for the benefit of wilderness. As new data sources come online, scientists and researchers will come up with uses for it and will develop the tools to develop hitherto impossible-to-obtain information. From this we will be able to create new knowledge, develop new theories and better hone our understanding of the importance of nature, ecology and wilderness.

People: Although our concern in this text is primarily with mapping wilderness, mapping people and getting people involved in the mapping process will be the key to developing a wider and more responsible value system focus not just on immediate human needs but that of the wider natural world and wilderness itself. Understanding the natural world and how our own survival is inextricably linked to the survival of wild nature and natural processes will shape future conservation efforts and, it is hoped, reverse some of the damage done in the past. People are at the centre of the conservation ethic and rewilding as it won't happen without our involvement. People are a huge data resource and crowdsourcing and other participatory approaches to data collection, validation and verification will augment the technological advances in sensors and mapping programmes.

Rewilding: This is perhaps the biggest challenge we face in terms of wilderness; namely, recognising what we have lost, being able to map it and then using that information and knowledge to target areas for putting it back. The ecological process of rewilding is not easy and is fraught with difficulties to do with getting the right species mix, creating appropriate habitats for reintroductions, developing connectivity between target areas to allow for species migration, etc. but making space for new wilderness within already managed landscapes is often controversial as it involves many conflicting interests. Nonetheless, the disbenefits to a few are most likely far outweighed by the benefits to the many in terms of delivery of ecosystem services, so interest in rewilding as a solution to a growing number of environmental problems is likely to develop rapidly over the next few years.

Policy: This leaves us with policy and associated legislation. They say that "a week is a long time in politics" and if true (which it generally is) this creates a bit of an awkward place for wilderness and rewilding as these simply do not fit easily on political timescales. Nevertheless, there is a growing interest in rewilding across the world which mirrors the similar growth in wilderness and other protected areas over the last century. Policy and legislation is in a constant state of flux, so we need longer term visions for wilderness protection and rewilding. The mapping work described here in this book can only serve to underpin and support this.

References

Aplet, G., Thomson, J., & Wilbert, M. (2000). Indicators of wildness: Using attributes of the land to assess the context of wilderness, Proceedings RMRS-P-15-VOL-2, *Wilderness science in a time of change conference-Volume 2: Wilderness within the context of larger systems*. Ogden: USDA Forest Service.

Asselen, S., & Verburg, P. H. (2013). Land cover change or land-use intensification: Simulating land system change with a global-scale land change model. *Global Change Biology, 19*(12), 3648–3667.

Carver, S., Comber, A., McMorran, R., & Nutter, S. (2012). A GIS model for mapping spatial patterns and distribution of wild land in Scotland. *Landscape and Urban Planning, 104*(3), 395–409.

Carver, S., Tricker, J., & Landres, P. (2013). Keeping it wild: Mapping wilderness character in the United States. *Journal of Environmental Management, 131*, 239–255.

Cholchester, M. (1996). Beyond "participation": Indigenous peoples, biological diversity conservation and protected area management, Unasyla. *An international journal of forestry and forest industries, 47*(3), 33–39.

Fisher, M., Carver, S., Kun, Z., McMorran, R., Arrell, K., & Mitchell, G. (2010). *Review of status and conservation of wild land in Europe,* Project commissioned by the Scottish government. Leeds: Wildland Research Institute.

Fletcher, C. (1971). *The Man who walked through time*. New York: A.A.Knopf.

Halpern, B. S., Walbridge, S., Selkoe, K. A., Kappel, C. V., Micheli, F., D'Agrosa, C., & Watson, R. (2008). A global map of human impact on marine ecosystems. *Science, 319*(5865), 948–952.

Havlik, P., Schneider, U. A., Schmid, E., Böttcher, H., Fritz, S., Skalský, R., & Obersteiner, M. (2011). Global land-use implications of first and second generation biofuel targets. *Energy Policy, 39*(10), 5690–5702.

Henry, D., & Husby, E. (1994). *Wilderness quality mapping in the Euro-Arctic Barents region: A potential management tool.* Trondheim: Direktoratet for Naturforvaltning, 1994. http://www.esri.com/recourses/userconf/proc95/to150/p113.html

Kuiters, A. T., van Eupen, M., Carver, S., Fisher, M., Kun, Z., & Vancura, V. (2013). *Wilderness register and indicator for Europe final report.* EEA Contract No: 07.0307/2011/610387/SER/B.3.

Leopold, A. (1949). *A sand county almanac: With other essays on conservation from Round River*. New York: OUP.

Sanderson, E. W., Jaiteh, M., Levy, M. A., Redford, K. H., Wannebo, A. V., & Woolmer, G. (2002). The human footprint and the last of the wild. *BioScience, 52*(10), 891–904.

Taylor, R. (Ed.). (2011). Chapter 1: Forests for a living planet, World Wide Fund for Nature, Gland, p. 35 WWF Living Forests Report.

Tricker, J., Landres, P., Dingman, S., Callagan, C., Stark, J., Bonstead, L., Fuhrmann, K., & Carver, S. (2012). *Mapping wilderness character in Death Valley National Park* (Natural resource report NPS/DEVA/NRR-2012/503). Fort Collins: National Park Service.

Index

S.J. Carver, S. Fritz (eds.), *Mapping Wilderness*,
DOI 10.1007/978-94-017-7399-7

MIX
Papier aus verantwortungsvollen Quellen
Paper from responsible sources
FSC® C105338

If you have any concerns about our products,
you can contact us on
ProductSafety@springernature.com

In case Publisher is established outside the EU,
the EU authorized representative is:
Springer Nature Customer Service Center GmbH
Europaplatz 3, 69115 Heidelberg, Germany

Printed by Libri Plureos GmbH
in Hamburg, Germany